Lecture Notes in Computer Science   1444
Edited by G. Goos, J. Hartmanis and J. van Leeuwen

Springer
Berlin
Heidelberg
New York
Barcelona
Budapest
Hong Kong
London
Milan
Paris
Singapore
Tokyo

Klaus Jansen   José Rolim   (Eds.)

Approximation Algorithms for Combinatorial Optimization

International Workshop APPROX'98
Aalborg, Denmark, July 18-19, 1998
Proceedings

 Springer

Series Editors

Gerhard Goos, Karlsruhe University, Germany
Juris Hartmanis, Cornell University, NY, USA
Jan van Leeuwen, Utrecht University, The Netherlands

Volume Editors

Klaus Jansen
IDSIA Lugano
Corso Elvezia 36, CH-6900 Lugano, Switzerland
E-mail: klaus@idsia.ch

José Rolim
University of Geneva, Computer Science Center
23, Rue Général Dufour, CH-1211 Geneva 4, Switzerland
E-mail: jose.rolim@cui.unige.ch

Cataloging-in-Publication data applied for

Die Deutsche Bibliothek - CIP-Einheitsaufnahme

Approximation algorithms for combinatorial optimization :
proceedings / International ICALP '98 Workshop, APPROX '98,
Aalborg, Denmark, July 18 - 19, 1998. Klaus Jansen ; José Rolim
(ed.). - Berlin ; Heidelberg ; New York ; Barcelona ; Budapest ; Hong
Kong ; London ; Milan ; Paris ; Singapore ; Tokyo : Springer, 1998
(Lecture notes in computer science ; Vol. 1444)
ISBN 3-540-64736-8

CR Subject Classification (1991): F.2.2, G.1.2, G.1.6, G.3, I.3.5

ISSN 0302-9743
ISBN 3-540-64736-8 Springer-Verlag Berlin Heidelberg New York

Typesetting: Camera-ready by author
SPIN 10638075   06/3142 – 5 4 3 2 1 0   Printed on acid-free paper

Preface

The Workshop on *Approximation Algorithms for Combinatorial Optimization Problems* **APPROX'98** focuses on algorithmic and complexity aspects arising in the development of efficient approximate solutions to computationally difficult problems. It aims, in particular, at fostering cooperation among algorithmic and complexity researchers in the field. The workshop, to be held at the University of Aalborg, Denmark, on July 18 - 19, 1998, co-locates with ICALP'98. We would like to thank the organizer of ICALP'98, Kim Larsen, for this opportunity. A previous event in Europe on approximate solutions of hard combinatorial problems consisting in a school followed by a workshop was held in Udine (Italy) in 1996.

Topics of interest for APPROX'98 are: design and analysis of approximation algorithms, inapproximability results, on-line problems, randomization techniques, average-case analysis, approximation classes, scheduling problems, routing and flow problems, coloring and partitioning, cuts and connectivity, packing and covering, geometric problems, network design, and various applications. The number of submitted papers to APPROX'98 was 37. Only 14 papers were selected. This volume contains the selected papers plus papers by invited speakers. All papers published in the workshop proceedings were selected by the program committee on the basis of referee reports. Each paper was reviewed by at least three referees who judged the papers for originality, quality, and consistency with the topics of the conference.

We would like to thank all authors who responded to the call for papers and our invited speakers: Magnús M. Halldórsson (Reykjavik), David B. Shmoys (Cornell), and Vijay V. Vazirani (Georgia Tech). Furthermore, we thank the members of the program committee:

- Ed Coffman (Murray Hill),
- Pierluigi Crescenzi (Florence),
- Ulrich Faigle (Enschede),
- Michel X. Goemans (Louvain and Cambridge),
- Peter Gritzmann (München),
- Magnús M. Halldórsson (Reykjavik),
- Johan Håstad (Stockholm),
- Klaus Jansen (Saarbrücken and Lugano, chair),
- Claire Kenyon (Orsay),
- Andrzej Lingas (Lund),
- George Lueker (Irvine),
- Ernst W. Mayr (München),
- Jose D.P. Rolim (Geneva, chair),
- Andreas Schulz (Berlin),
- David B. Shmoys (Cornell),
- Jan van Leeuwen (Utrecht).

and the reviewers Susanne Albers, Abdel-Krim Amoura, Gunnar Andersson, Christer Berg, Ioannis Caragiannis, Dietmar Cieslik, A. Clementi, Artur Czumaj, Elias Dahlhaus, A. Del Lungo, Martin Dyer, Lars Engebretsen, Thomas Erlebach, Uriel Feige, Stefan Felsner, Rudolf Fleischer, Andras Frank, R. Grossi, Joachim Gudmundsson, Dagmar Handke, Stephan Hartmann, Dorit S. Hochbaum, J.A. Hoogeveen, Sandra Irani, Jesper Jansson, Mark Jerrum, David Johnson, Christos Kaklamanis, Hans Kellerer, Samir Khuller, Ekkehard Koehler, Stefano Leonardi, Joseph S. B. Mitchell, Rolf H. Möhring, S. Muthu Muthukrishnan, Petra Mutzel, Giuseppe Persiano, Joerg Rambau, Ramamoorthi Ravi, Ingo Schiermeyer, Martin Skutella, Roberto Solis-Oba , Frederik Stork, Ewald Speckenmeyer, C.R. Subramanian, Luca Trevisan, Denis Trystram, John Tsitsiklis, Marc Uetz, Hans-Christoph Wirth, Gerhard Woeginger, Martin Wolff, Alexander Zelikovsky, and Uri Zwick.

We gratefully acknowledge sponsorship from the Max-Planck-Institute for Computer Science Saarbrücken (AG 1, Prof. Mehlhorn), ALCOM-IT Algorithms and Complexity in Information Technology, and Siemens GmbH. We also thank Luca Gambardella, the research institute IDSIA Lugano, Alfred Hofmann, Anna Kramer, and Springer-Verlag for supporting our project.

May 1998   Klaus Jansen

Contents

Approximations of Independent Sets in Graphs

Magnús M. Halldórsson[1,2]

[1] Science Institute, University of Iceland, Reykjavik, Iceland. mmh@hi.is
[2] Department of Informatics, University of Bergen, Norway.

1   Introduction

The *independent set problem* is that of finding a maximum size set of mutually non-adjacent vertices in a graph. The study of independent sets, and their *alter egos*, cliques, has had a central place in combinatorial theory.

Independent sets occur whenever we seek sets of items free of pairwise conflicts, e.g. when scheduling tasks. Aside from numerous applications (which might be more pronounced if the problems weren't so intractable), independent sets and cliques appear frequently in the theory of computing, e.g. in interactive proof systems [6] or monotone circuit complexity [2]. They form the representative problems for the class of subgraph or packing problems in graphs, are essential companions of graph colorings, and form the basis of clustering, whether in terms of nearness or dispersion.

As late as 1990, the literature on independent set approximations was extremely sparse. In the period since Johnson [31] started the study of algorithms with good *performance ratios* in 1974 – and in particular showed that a whole slew of independent set algorithms had only the trivial performance ratio of n on general graphs – only one paper had appeared containing positive results [29], aside from the special case of planar graphs [34, 8]. Lower bounds were effectively non-existent, as while it was known that the best possible performance ratio would not be some *fixed* constant, there might still be a polynomial-time approximation scheme lurking somewhere.

Success on proving lower bounds for Independent Set has been dramatic and received worldwide attention, including the New York Times. Progress on improved approximation algorithms has been less dramatic, but a notable body of results has been developed. The purpose of this talk is to bring some of these results together, consider the lessons learned, and hypothesize about possible future developments.

The current paper is not meant to be the ultimate summary of independent set approximation algorithms, but an introduction to the performance ratios known, the strategies that have been applied, and offer glimpses of some of the results that have been proven.

We prefer to study a range of algorithms, rather than seek only the best possible performance guarantee. The latter is fine as far as it goes, but is not the only thing that matters; only so much information is represented by a single number. Algorithmic strategies vary in their time requirements, temporal access to data, parallelizability, simplicity and numerous other factors that are far from irrelevant. Different algorithms may also be incomparable on different classes of graphs, e.g. depending on the size of the optimal solution. Finally, the proof techniques are perhaps the most valuable product of the analysis of heuristics.

We look at a slightly random selection of approximation results in the body of the paper. A complete survey is beyond the scope of this paper but is under preparation. The primary criteria for selection was simplicity, of the algorithm and the proof. We state some observations that have not formally appeared before, give some recent results, and present simpler proofs of other results.

The paper is organized as follows. We define relevant problems and definitions in the following section. In the body of the paper we present a number of particular results illustrating particular algorithmic strategies: subgraph removal, semi-definite programming, partitioning, greedy algorithms and local search. We give a listing of known performance results and finish with a discussion of open issues.

2 Problems and definitions

INDEPENDENT SET: Given a graph $G = (V, E)$, find a maximum cardinality set $I \subseteq V$ such that for each $u, v \in I$, $(u, v) \notin E$. The *independence number* of G, denoted by $\alpha(G)$, is the size of the maximum independent set.

CLIQUE PARTITION: Given a graph $G = (V, E)$, find a minimum cardinality set of disjoint cliques from G that contains every vertex.

κ-SET PACKING: Given a collection C of sets of size at most κ drawn from a finite set S, find a minimum cardinality collection C' such that each element in S is contained in some set in C'.

These problems may also be weighted, with weights on the vertices (or on the sets in SET PACKING).

A set packing instance is a case of an independent set problem. Given a set system (C, S), form a graph with a vertex for each set in C and edge between two vertices if the corresponding sets intersect. Observe that if the sets in C are of size at most κ, then the graph contains a $\kappa + 1$-*claw*, which is a subgraph consisting of a center node adjacent to $\kappa + 1$ mutually non-adjacent vertices. The independent set problem in $\kappa + 1$-claw free graphs slightly generalizes κ-SET PACKING, which in turn slightly generalizes κ-DIMENSIONAL MATCHING.

The *performance ratio* ρ_A of an independent set algorithm A is given by

$$\rho_A = \rho_A(n) = \max_{G, |G|=n} \frac{\alpha(G)}{A(G)}.$$

Notation

n	the number of vertices	$d(v)$	the degree of vertex v
m	the number of edges	$N(v)$	set of neighbors of v
Δ	maximum degree	$\overline{N}(v)$	non-neighbors of v
$\bar{d}$	average degree	$A(G)$	the size of solution found by A
δ	minimum degree	ρ_A	performance ratio of A
α	independence number	$\overline{\chi}$	clique partition number
κ	maximum claw size		

3 Ramsey theory and subgraph removal

The first published algorithm with a non-trivial performance ratio on general graphs was introduced in 1990. In appreciation of the heritage that the late master Erdős left us, we give here a treatment different from Boppana and Halldórsson [12] that more closely resembles the original Ramsey theorem of Erdős and Szekeres [17].

Ramsey (G)
  if $G = \emptyset$ then return $(\emptyset, \emptyset)$
  choose some $v \in G$
  $(C_1, I_1) \leftarrow$ Ramsey$(N(v))$
  $(C_2, I_2) \leftarrow$ Ramsey$(\overline{N}(v))$
  return (larger of $(C_1 \cup \{v\}, C_2)$,
        larger of $(I_1, I_2 \cup \{v\})$)

CliqueRemoval (G)
  $i \leftarrow 1$
  $(C_i, I_i) \leftarrow$ Ramsey (G)
  while $G \neq \emptyset$ do
    $G \leftarrow G - C_i$
    $i \leftarrow i + 1$
    $(C_i, I_i) \leftarrow$ Ramsey (G)
  od
  return $((\max_{j=1}^{i} I_j), \{C_1, C_2, \ldots, C_i\})$

Fig. 1. Independent set algorithm based on Ramsey theory

Theorem 1. Ramsey *finds an independent set* I *and a clique* C *such that* $\binom{|I|+|C|}{|C|} - 1 \geq n$. *In particular,* $|I| \cdot |C| \geq \frac{1}{4} \log^2 n$.

Proof. The proof is by induction on both $|I|$ and $|C|$. It is easy to verify the claim when either $|I|$ or $|C|$ are at most 1. By the induction hypothesis,

$$n = |N(v)| + |\overline{N(v)}| + 1 \leq \left(\binom{|I_1| + |C_1|}{|C_1|} - 1\right) + \left(\binom{|I_2| + |C_2| - 1}{|C_2|} - 1\right) + 1.$$

Recall that $|C| = \max(|C_1| + 1, |C_2|)$ and $|I| = \max(|I_1|, |I_2| + 1)$. Thus,

$$n \leq \binom{|I| + |C| - 1}{|C| - 1} + \binom{|I| + |C| - 1}{|C|} - 1.$$

The claim now follows from the equality $\binom{s+t}{s} = \binom{s+t-1}{t-1} + \binom{s+t-1}{t}$.

It is easy to verify that the product of $|I|$ and $|C|$ is minimized when they are equal. That is, when $n = \binom{2|C|}{|C|} \leq 2^{2|C|}$, hence $|I| \cdot |C| \geq (\frac{1}{2} \log n)^2$. ∎

The following simplified proof of a $O(n/\log^2 n)$ performance ratio also borrows from another of Erdős's work [15].

Theorem 2. *The performance ratio of* CliqueRemoval *is* $O(n/\log^2 n)$.

Proof. Let CC denote the number of cliques returned by CliqueRemoval and let CC_0 denote the number of cliques removed before the size of the graph dropped below $n_0 = n/\log^2 n$. Let t be the size of the smallest of these latter cliques, which without loss of generality is at most $\log^2 n$. Then $CC_1 \leq n/t$, and $CC \leq n/t + n_0 \leq 2n/t$.

If I is the independent set returned, we have that $|I| \geq 4\log^2 n_0/t \geq 2\log^2 n/t$. Consider the product of the two performance ratio of CliqueRemoval, ρ_α for independent sets, and $\rho_{\overline{\chi}}$ for clique partition:

$$\rho_\alpha \cdot \rho_{\overline{\chi}} = \frac{CC}{\overline{\chi}} \cdot \frac{\alpha}{|I|} \leq \frac{n}{\log^2 n} \frac{\alpha}{\overline{\chi}} \leq \frac{n}{\log^2 n}.$$

Clearly, either performance ratio is also bounded by $O(n/\log^2 n)$. ∎

For graphs with high independence number, the ratios are better.

Theorem 3. *If* $\alpha(G) \geq n/k + p$, *then* CliqueRemoval *finds an independent set of size* $\Omega(p^{1/(k-1)})$.

This illustrates the strategy of *subgraph removal*, that is based around the concept that graphs without small dense subgraphs are easier to approximate.

4 Lovász theta function

A fascinating polynomial-time computable function $\vartheta(G)$, that was introduced by Lovász [37], has the remarkable *sandwiching* property that it always lies between two NP-hard functions, $\alpha(G) \leq \vartheta(G) \leq \overline{\chi}(G)$. This property suggests that it may be particularly suited for obtaining good approximations to either function. While some of those hopes have been dashed, a number of fruitful applications have been found and it remains the most promising candidate for obtaining improved approximations.

Karger, Motwani and Sudan [32] proved the following property in the context of coloring. The "soft-omega" notation $\tilde{\Omega}$ hides logarithmic factors.

Theorem 4 (Karger et al). *If* $\vartheta(\overline{G}) \leq k$, *then an independent set of size* $\tilde{\Omega}(n^{3/(k+1)})$ *can be constructed with high probability in polynomial time.*

Mahajan and Ramesh [38] showed how these and related algorithms can be derandomized. Alon and Kahale [4] applied the theta function further for independent sets.

Theorem 5 (Alon, Kahale). *If $\vartheta(G) \geq n/k+p$ (e.g. if $\alpha(G) \geq n/k+p$), then we can find a graph K on p vertices with $\vartheta(\overline{K}) \leq k$.*

Combining the two, they obtained a ratio for high-independence graphs that improves on Theorem 3.

Corollary 1. *For any fixed integer $k \geq 3$, if $\vartheta(G) \geq n/k + p$, then an independent set of size $\tilde{\Omega}(p^{3/(k+1)})$ can be found in polynomial time.*

Theta function on sparse graphs Karger et al. proved a core result in terms of maximum degree of the graph. In fact, their argument also holds in terms of average degree.

Theorem 6 (Karger et al). *If $\vartheta(\overline{G}) \leq k$, then an independent set of size $\tilde{\Omega}(n/\overline{d}^{1-2/k})$ can be constructed with high probability in polynomial time.*

Vishwanathan [40] observed that this, combined with Theorem 5, also yields an improved algorithm for bounded-degree graphs. This, however, has not been stated before in the literature, to the best of the author's knowledge.

Proposition 1. INDEPENDENT SET *can be approximated within a factor of* $O(\Delta \log \log \Delta / \log \Delta)^1$.

Proof. Given G, if $\alpha(G) \geq n/k = n/2k + n/2k$, then we can find a subgraph K on $n/2k$ vertices with $\vartheta(\overline{K}) \leq k$ and maximum degree at most $\Delta(G)$, by Theorem 5. By Theorem 6, we can find an independent set in K (and G) of size $\Omega((n/2k)/\Delta^{1-2/k}) = \Omega(n/\Delta \cdot \Delta^{2/k}/k)$. If $k \leq \log \Delta / \log \log \Delta$, then the set found is of size $\Omega(n/\Delta \cdot \log \Delta)$, and the claim is satisfied since $\alpha \leq n$. Otherwise, $\alpha \leq n \log \log \Delta / \log \Delta$, and any maximal solution is of size $n/(\Delta+1)$, for a ratio satisfying the proposition. ∎

5 Partitioning and weighted independent sets

A simple strategy in the design of approximation algorithms is to break the problem into a collection of easier subproblems.

Observation 1 *Suppose we can partition G into t subgraphs and solve the weighted independent set problem on each subgraph optimally. Then, the largest of these solutions is a t-approximation of G.*

Proof. The size of the optimal solution for G is at most the sum of the sizes of the largest independent sets on each subgraph, which is at most t times the largest solution in some subgraph. ∎

This gives us the first non-trivial ratio for weighted independent sets in general graphs [21].

---

[1] We may need to assume that Δ be a large constant independent of n.

Theorem 7. *The weighted independent set problem can be approximated within* $O(n(\log\log n/\log n)^2)$.

Proof. The bound on Ramsey in Theorem 1 implies that it either outputs an independent set of size $\log^2 n$, or a clique of size $\log n/\log\log n$. We apply this algorithm repeatedly, like CliqueRemoval, but either removes a $\log^2 n$-independent set or a $\log n/\log\log n$-clique in each step.

We now form a partition where each class is either an independent set, or a (not necessarily disjoint) union of $\log n/\log\log n$ different cliques. This yields a partition into $O(n(\log\log n/\log n)^2)$ classes.

The weighted independent set problem on such classes can be solved by exhaustively checking all $(\log n/\log\log n)^{\log n/\log\log n} = O(n)$ possible combinations of selecting one vertex from each clique. Thus, by the above observation, the claimed ratio follows. ∎

On bounded-degree graphs, we can apply a partitioning lemma of Lovász [35], which we specialize here to this application.

Lemma 2. *The vertices of a graph can be partitioned into* $\lceil(\Delta + 1)/3\rceil$ *sets, where each induces a subgraph of maximum degree at most two.*

Proof. Start with an arbitrary partitioning into $\lceil(\Delta + 1)/3\rceil$ sets, and repeat the following operation: If v is adjacent to three or more vertices in its set, move it to a set where it has at most two neighbors. Such a set must exist as otherwise v's degree would be at least $3\lceil(\Delta + 1)/3\rceil \geq \Delta + 1$. Observe that such a move increases the number of *cross edges*, or edges going between different sets, hence this process must terminate with a partition where every vertex has at most two neighbors in its set. ∎

Dynamic programming easily solves the weighted maximum independent set problem on each such subgraph, and as shown in [23], the partitioning can also be performed in linear time by starting with a greedy partition.

Theorem 8. WEIGHTED INDEPENDENT SET *is approximable within* $\lceil(\Delta+1)/3\rceil$ *in linear time.*

Hochbaum [29] also used a form of a partition, a coloring, to approximate weighted independent set problems.

6 Greediness and Set packing

The general set packing problem can be shown to be equivalent to the independent set problem. Given a graph $G = (V, E)$, let the base set S contain one element for each edge in E, and for each vertex $v \in V$, form a set C_v containing the base sets corresponding to edges incident on v. It holds that the maximum number of sets in a set packing of $(S, \mathcal{C})$ is $\alpha(G)$.

There are four parameters of set systems that are of interest for SET PACKING approximations: n, the number of sets, $|S|$, the number of base elements, κ, maximum cardinality of a set, and B, the maximum number of occurrences of a base elements in sets in C.

In the reduction above, we find that $|C| = n$, and therefore approximations of Independent Set as functions of n carry over to approximations of Set Packing in terms of $|C|$. A reduction in the other direction also preserves this relationship.

As for B, observe that in the reduction above, $B = 2$, for arbitrary instances. Hence, we cannot expect any approximations as functions of B alone. It remains to consider approximability in terms of κ and $|S|$.

Local search Just about any solution gives a modest approximation.

Theorem 9. *Any maximal solution is κ-approximate.*

Proof. We say that a vertex v *dominates* a vertex u if u and v are either adjacent or the same vertex. Any vertex can dominate at most κ vertices from an optimal solution. Yet, maximality requires that a maximal independent set dominates all vertices of the graph. Hence, an optimal solution is at most κ times bigger than any maximal solution. ∎

This can be strengthened using simple local search. Tight analysis was first given by Hurkens and Schrijver [30], whose article title seemed to obscure its contents since the results were reproduced in part or full by several groups of authors [33, 42, 20, 43, 7].

Local search is straightforward for problems whose solutions are collections of items: repeatedly try to extend the solution by eliminating t elements while adding $t + 1$ elements. A solution that cannot be further extended by such improvements is said to be *t-optimal*. It turns out that 2-optimal solutions, which are the most efficient and the easiest to analyze, already give considerably improved approximations.

Theorem 10. *Any 2-optimal solution is $(\kappa + 1)/2$-approximate.*

Proof. Let us argue in terms of independent sets in $\kappa + 1$-claw free graphs. Let I be a 2-optimal solution, and let O be any optimal independent set. Partition O into O_1, those vertices in O that are adjacent to only one vertex in I, and O_2, those vertices in O that are adjacent to two or more vertices in I. Note that each vertex in I is adjacent to at most κ vertices in O, due to the lack of a $\kappa + 1$-claw. Then, considering the edges between I and O we have that

$$|O_1| + 2|O_2| \le \kappa|I|.$$

Also, since I is 2-optimal

$$|O_1| \le |I|.$$

Adding the two inequalities gives that

$$2|O| = 2(|O_1| + |O_2|) \le (\kappa + 1)|I|,$$

or that the performance ratio is at most $(\kappa + 1)/2$. ∎

Using t-opt we can prove a bound of $\kappa/2+\epsilon$. On bounded-degree graphs, local search can be applied with a very large radius while remaining in polynomial time, and using some additional techniques, Berman, Fürer, and Fujito [11, 10] obtained the best performance ratios known for small values of Δ of $(\Delta + 3)/5$.

Greedy algorithms This leaves $|S|$ as the only parameter left to be studied for SET PACKING.

A related topic is the *Strong Stable Set* problem, where we seek an independent set in which the vertices are of distance at least two apart. Such a strong stable set corresponds to a set packing of the set system formed by the closed vertex neighborhoods in the graph. In this case, $C = |S| = n$. The question is then whether this is easier to approximate than the general independent set problem.

Halldórsson, Kratochvíl, and Telle [22] recently gave a simple answer to this question, using a greedy set packing algorithm that always picks the smallest set remaining.

Theorem 3 *Set Packing can be approximated within $\sqrt{|S|}$ in time linear in the input size.*

Proof. Consider the following greedy algorithm. In each step, it chooses a smallest set and removes from the collection all sets containing elements from the selected set.

```
GreedySP(S,C)
   t ← 0
   repeat
      t ← t + 1
      Xt ← C ∈ C of minimum cardinality
      Zt ← {C ∈ C : X ∩ C ≠ ∅}
      C ← C - Zt
   until |C| = 0
   Output {X1, X2, ..., Xt}
```

Let $M = \lfloor\sqrt{|S|}\rfloor$. Observe that $\{Z_1, \ldots, Z_t\}$ forms a partition of C. Let i be the index of some iteration of the algorithm, i.e. $1 \leq i \leq t$. All sets in Z_i contain at least one element of X_i, thus the maximum number of disjoint sets in Z_i is at most the cardinality of X_i. On the other hand, every set in Z_i is of size at least X_i, so the maximum number of disjoint sets in Z_i is also at most $\lfloor|S|/|X_i|\rfloor$. Thus, the optimal solution contains at most $\min(|X_i|, \lfloor|S|/|X_i|\rfloor) \leq \max_x \min(x, \lfloor|S|/x\rfloor) = M$ sets from Z_i.

Thus, in total, the optimal solution contains at most tM sets, when the algorithm finds t sets, for a ratio of at most M. ∎

Observe that this approximation is near the best possible. Since a graph contains $O(n^2)$ edges, Håstad's result [27] yields an $\Omega(m^{1/2-\epsilon})$ lower bound, for any $\epsilon > 0$.

Other greedy algorithms have been studied, especially the one that repeatedly selects vertices of minimum degree in the graph. It remains, e.g., the driving force for the best ratio known for sparse graphs, $(2\bar{d} + 3)/5$ [26].

7 Summary of results

Table 1 contains a listing of the various ratios that have been proved for heuristics for the independent set problem, along with known inapproximability results. It is divided according to graph classes / graph parameter. Results in terms of other measures of graphs or pairs of measures are not included. The results hold for unweighted graphs except for the last category. Each entry contains the ratio proved, the algorithmic strategy used, the complexity of the method, and a citation.

We have not described the Nemhauser-Trotter reduction [39] that was championed by Hochbaum [29], which allows one to assume in many cases without loss of generality that the maximum weight independent set is of weight at most half the total weight of the graph. The complexity of this procedure equals the complexity of finding a minimum cut in a network in the weighted case ($O(nm)$), and the complexity of bipartite matching in the unweighted case ($O(\sqrt{n}m)$).

Abbreviations: SR = subgraph removal, SDP = semi-definite programming, NT = Nemhauser-Trotter reduction, MIS = arbitrary maximal independent set.

Complexity: NT refers to the complexity of the Nemhauser-Trotter reduction. "Linear" means time linear in the size of the graph. $n^{O(1)}$ suggests time bounded by a polynomial of high degree; in the case of the $(\Delta + 3)/5$ ratio, the degree of the polynomial appears to be on the order of 2^{100} [24].

8 Discussion

A number of open issues remain.

General graphs There remains some gap between the best upper and lower bounds known for general graphs. Stated in terms of "distance from trivial", it is the difference between $\log^2 n$ and $n^{o(1)}$. It is not as presumptuous now to conjecture that the ultimate ratio is $n/polylog(n)$ as it was in 1991 [19]. It may be possible to extend the proof of [27] to argue a stronger lower bound than $n^{1-\epsilon}$ if given a stronger assumption, such as SAT not having $2^{o(n)}$ time algorithms. (Admittedly, such a task appears less than trivial [28]).

Performance of ϑ-function The theta function remains the most promising candidate for improved approximations. Some of the hopes attached with it have been dashed. Feige [18] showed that its performance ratio is at least $n/2^{O(\sqrt{\log n})}$. Can't we at least prove something better than the simple Ramsey-theoretic bound?

High-independence graphs Gaps in bounds on approximability are nowhere greater than in the case of independent sets in graphs with $\alpha(G) = n/k$, for some fixed $k > 2$. These problems are APX-hard, i.e. hard within some

Result	Method	Complexity	Reference
General graphs			
$O(n/\log n)$		$O(nm)$	[41]
$O(n/\log^2 n)$	SR	$O(nm)$	[12]
$\Omega(n^{1-\epsilon})$			[27]
High-independence graphs ($\alpha = n/k$)			
$O(n^{1-1/(k-1)})$	SR	$O(nm)$	[12]
$\tilde{O}(n^{1-3/(k+1)})$	SDP	SDP	[4]
$\Omega(1+c)$			[6]
Sparse graphs			
$\bar{d}+1$	Greedy	linear	[29], via [16]
$(\bar{d}+2)/2$	Greedy	linear	[26]
$(\bar{d}+1)/2$	Greedy + NT	NT	[29]
$(2\bar{d}+4.5)/5$	Greedy+SR	linear	[25]
$(2\bar{d}+3)/5$	Greedy + NT	NT	[26]
Bounded-degree graphs			
Δ	MIS	linear	
$\Delta/2$	Brooks+NT	NT	[29], via [36]
$(\Delta+2)/3$	Greedy	linear	[26]
$(\Delta+3)/5$	Local search +	$n^{O(1)}$	[11, 10]
$(\Delta+2)/4+\epsilon$	Local search	$\Delta^{O(\Delta)}n$	[24]
$\Delta/6+O(1)$	SR	$O(\Delta^{\Delta}n+n^2)$	[25]
$O(\Delta/\log\log\Delta)$	SR	$n^{O(1)}$	[25], via [1]
$O(\Delta\log\log\Delta/\log\Delta)$	SDP	SDP	[40], via [32, 4]
$\Omega(\Delta^c)$			[3]
$\kappa + 1$-claw-free graphs and Set Packing			
κ	MIS	linear	
$(\kappa+1)/2$	Local search	$O(n^3)$	[33, 42]
$\kappa/2+\epsilon$	Local search	$O(n^{\log_\kappa 1/\epsilon})$	[30, 20, 43]
$\sqrt{|S|}$	GreedySP	linear	[22]
$\Omega(\kappa^c)$			[3]
$\Omega(|S|^{1-\epsilon})$			[27]
Weighted graphs			
$\Delta/2$	Brooks+NT	NT	[29]
$\lceil(\Delta+1)/3\rceil$	Partitioning	linear	[23], via [35]
$(\Delta+2)/3$	Partitioning+NT	NT	[23]
κ	Max-weight greedy	$O(n^2)$	[29]
$\kappa-1+\epsilon$	Local search	$n^{O(1/\epsilon)}$	[5, 7]
$(4\kappa+2)/5$	LS + greedy	$n^{O(\kappa)}$	[13]
$O(n(\log\log n/\log n)^2)$	SR+Partitioning	$O(n^2)$	[21]

Table 1. Results on approximating independent sets

constant factor greater than one, but all the upper bounds known are some roots of n. These problems generalize the case of k-colorable graphs, for which a similar situation holds. Results of Alon and Kahale [4] indicate that some root of n is also the best that the theta function will yield in this case. The limited progress on the more studied k-coloring problem suggests that this is near best possible.

Vertex cover The preceding item has relevance to the approximability of VERTEX COVER, which is the problem of finding a minimum set of vertices S such that $V - S$ is an independent set. If VERTEX COVER can be approximated within less than 1.5, then INDEPENDENT SET in graphs with $\alpha = n/3$ is constant approximable and GRAPH 3-COLORING is $O(\log n)$ approximable, as first shown by Bar-Yehuda and Moran [9]. This gives support to the conjecture that factor 2 is optimal for VERTEX COVER, within lower order terms [29].

Bounded-degree graphs It is natural to extrapolate that the improved hardness ratio $n^{1-\epsilon}$ of [27] indicates that the hardness ratio $\Omega(\Delta^c)$ of [3] for bounded-degree graphs could be jacked up to $\Omega(\Delta^{1-o(1)})$.
From the upper bound side, it would be nice to extend the $o(\Delta)$ ratios of [25, 40] to hold for all values of Δ as a function of n. Demange and Paschos [14] have parametrized the strategy of [25] to give a ratio Δ/c for every c, that holds for every value of Δ in time $O(n^c)$.

$\kappa + 1$-claw-free graphs Claw-free graphs appear considerably harder than bounded-degree graphs. Any improvement to the $\kappa/2 + \epsilon$ ratios would be most interesting. Observe that a maximum κ-set packing is within a factor κ from a *minimum hitting set* of a collection of sets of size κ, but we also do not have any better ratio than factor κ for the latter problem.
In the weighted case, we know that the greedy and the local search strategies do not improve on $\kappa - O(1)$. However, the combination of the two does attain asymptotically better ratios [13]. We conjecture that selective local search starting from a greedy solution does attain the unweighted bound of $\kappa/2 + \epsilon$.

$o(\alpha)$-approximations While we do have $o(n)$-approximations of INDEPENDENT SET, these methods fail to give us anything beyond the trivial when, say, $\alpha = \sqrt{n}$. While it is probably too much to ask for a $\omega(1)$-size independent set in graphs with $\alpha \approx \log n$, it is not unfair to ask for, say, a $\alpha \log n / \log^2 \alpha$-approximation.

References

1. M. Ajtai, P. Erdős, J. Komlós, and E. Szemerédi. On Turán's theorem for sparse graphs. *Combinatorica*, 1(4):313–317, 1981.
2. N. Alon and R. B. Boppana. The monotone complexity of Boolean functions. *Combinatorica*, 7(1):1–22, 1987.
3. N. Alon, U. Feige, A. Wigderson, and D. Zuckerman. Derandomized graph products. *Computational Complexity*, 5(1):60 – 75, 1995.
4. N. Alon and N. Kahale. Approximating the independence number via the θ function. *Math. Programming*. To appear.

5. E. M. Arkin and R. Hassin. On local search for weighted k-set packing. *ESA '97*, LNCS 1284.
6. S. Arora, C. Lund, R. Motwani, M. Sudan, and M. Szegedy. Proof verification and hardness of approximation problems. *FOCS '92*, 14–23.
7. V. Bafna, B. O. Narayanan, and R. Ravi. Non-overlapping local alignments (weighted independent sets of axis parallel rectangles). *WADS '95*, LNCS 955, 506–517.
8. B. S. Baker. Approximation algorithms for NP-complete problems on planar graphs. *J. ACM*, 41:153–180, Jan. 1994.
9. R. Bar-Yehuda and S. Moran. On approximation problems related to the independent set and vertex cover problems. *Discrete Appl. Math.*, 9:1–10, 1984.
10. P. Berman and T. Fujito. On the approximation properties of independent set problem in degree 3 graphs. *WADS '95*, LNCS 955, 449–460.
11. P. Berman and M. Fürer. Approximating maximum independent set in bounded degree graphs. *SODA '94*, 365–371.
12. R. B. Boppana and M. M. Halldórsson. Approximating maximum independent sets by excluding subgraphs. *BIT*, 32(2):180–196, June 1992.
13. B. Chandra and M. M. Halldórsson. Approximating weighted k-set-packing. Manuscript, May 1998.
14. M. Demange and V. T. Paschos. Improved approximations for maximum independent set via approximation chains. *Appl. Math. Lett*, 1996. To appear.
15. P. Erdős. Some remarks on chromatic graphs. *Colloq. Math.*, 16:253–256, 1967.
16. P. Erdős. On the graph theorem of Turán (in Hungarian). *Mat. Lapok*, 21:249–251, 1970.
17. P. Erdős and G. Szekeres. A combinatorial problem in geometry. *Compositio Math.*, 2:463–470, 1935.
18. U. Feige. Randomized graph products, chromatic numbers, and the Lovász ϑ-function. *Combinatorica*, 17(1):79–90, 1997.
19. M. M. Halldórsson. A still better performance guarantee for approximate graph coloring. *Inform. Process. Lett.*, 45:19–23, 25 January 1993.
20. M. M. Halldórsson. Approximating discrete collections via local improvements. *SODA '95*, 160–169.
21. M. M. Halldórsson. Approximation via partitioning. Res. Report IS-RR-95-0003F, Japan Adv. Inst. of Sci. and Tech., Mar. 1995.
22. M. M. Halldórsson, J. Kratochvíl, and J. A. Telle. Independent sets with domination constraints. *ICALP '98*, LNCS.
23. M. M. Halldórsson and H. C. Lau. Low-degree graph partitioning via local search with applications to constraint satisfaction, max cut, and 3-coloring. *J. Graph Algo. Applic.*, 1(3):1–13, 1997.
24. M. M. Halldórsson and J. Radhakrishnan. Improved approximations of independent sets in bounded-degree graphs. *SWAT '94*, LNCS 824, 195–206.
25. M. M. Halldórsson and J. Radhakrishnan. Improved approximations of independent sets in bounded-degree via subgraph removal. *Nordic J. Computing*, 1(4):475–492, 1994.
26. M. M. Halldórsson and J. Radhakrishnan. Greed is good: Approximating independent sets in sparse and bounded-degree graphs. *Algorithmica*, 18:145–163, 1997.
27. J. Håstad. Clique is hard to approximate within $n^{1-\epsilon}$. *FOCS '96*, 627–636.
28. J. Håstad. Private communication, 1997.
29. D. S. Hochbaum. Efficient bounds for the stable set, vertex cover, and set packing problems. *Disc. Applied Math.*, 6:243–254, 1983.

30. C. A. J. Hurkens and A. Schrijver. On the size of systems of sets every t of which have an SDR, with an application to the worst-case ratio of heuristics for packing problems. *SIAM J. Disc. Math.*, 2(1):68–72, Feb. 1989.

31. D. S. Johnson. Approximation algorithms for combinatorial problems. *J. Comput. Syst. Sci.*, 9:256–278, 1974.

32. D. Karger, R. Motwani, and M. Sudan. Approximate graph coloring by semidefinite programming. *FOCS '94*, 2–13.

33. S. Khanna, R. Motwani, M. Sudan, and U. Vazirani. On syntactic versus computational views of approximability. *FOCS '94*, 819–830.

34. R. J. Lipton and R. E. Tarjan. Applications of a planar separator theorem. *FOCS '77*, 162–170.

35. L. Lovász. On decomposition of graphs. *Stud. Sci. Math. Hung.*, 1:237–238, 1966.

36. L. Lovász. Three short proofs in graph theory. *J. Combin. Theory Ser. B*, 19:269–271, 1975.

37. L. Lovász. On the Shannon capacity of a graph. *IEEE Trans. Inform. Theory*, IT-25(1):1–7, Jan. 1979.

38. S. Mahajan and H. Ramesh. Derandomizing semidefinite programming based approximation algorithms. *FOCS '95*, 162–169.

39. G. L. Nemhauser and L. Trotter. Vertex packings: Structural properties and algorithms. *Math. Programming*, 8:232–248, 1975.

40. S. Vishwanathan. Personal communication, 1996.

41. A. Wigderson. Improving the performance guarantee for approximate graph coloring. *J. ACM*, 30(4):729–735, 1983.

42. G. Yu and O. Goldschmidt. On locally optimal independent sets and vertex covers. Manuscript, 1993.

43. G. Yu and O. Goldschmidt. Local optimality and its application on independent sets for k-claw free graphs. Manuscript, 1994.

Using Linear Programming in the Design and Analysis of Approximation Algorithms: Two Illustrative Problems

David B. Shmoys

Cornell University, Ithaca NY 14853, USA

Abstract. One of the foremost techniques in the design and analysis of approximation algorithms is to round the optimal solution to a linear programming relaxation in order to compute a near-optimal solution to the problem at hand. We shall survey recent work in this vein for two particular problems: the uncapacitated facility location problem and the problem of scheduling precedence-constrained jobs on one machine so as to minimize a weighted average of their completion times.

1 Introduction

One of the most successful techniques in the design and analysis of approximation algorithms for combinatorial optimization problems has been to first solve a relaxation of the problem, and then to round the optimal solution to the relaxation to obtain a near-optimal solution for the original problem. Although the relaxation used varies from problem to problem, linear programming relaxations have provided the basis for approximation algorithms for a wide variety of problems. Throughout this paper, we shall discuss approximation algorithms, where a *ρ-approximation algorithm* for an optimization problem is a polynomial-time algorithm that is guaranteed to find a feasible solution for the problem with objective function value within a factor of ρ of optimal.

In this brief survey, we shall discuss recent developments in the design of approximation algorithms for two specific problems, the uncapacitated facility location problem, and a rather basic single-machine scheduling problem. In focusing on just two problems, clearly we are omitting a great deal of important recent work on a wide cross-section of other problems, but the reader can obtain an accurate indication of the level of activity in this area by considering, for example, the other papers in this proceedings. For a more comprehensive review of the use of this approach, the reader is referred to the volume edited by Hochbaum [16].

We shall consider the following scheduling problem. There are n jobs to be scheduled on a single machine, where each job j has a specified weight w_j and processing time p_j, $j = 1, \ldots, n$, which we restrict to be positive integers. Furthermore, there is a partial order $\prec$ that specifies a precedence relation among the jobs; if $j \prec k$ then we must find a schedule in which job j completes its processing before job k is started. Each job must be processed without interruption, and the machine can process at most one job at a time. If we let C_j denote the completion time of job j, then we wish to minimize the average weighted completion time $\sum_{j=1}^{n} w_j C_j/n$, or equivalently, $\sum_{j=1}^{n} w_j C_j$. In the notation of Graham, Lawler, Lenstra, & Rinnooy Kan [11], the problem is denoted $1|prec| \sum w_j C_j$; it was shown to be $\mathcal{NP}$-hard by Lawler [21].

The first non-trivial approximation algorithm for $1|prec| \sum w_j C_j$ is due to Ravi, Agrawal, & Klein [33], who gave an $O(\lg n \lg W)$-approximation algorithm, where $W = \sum_j w_j$. A slightly improved performance guarantee of $O(\lg n \lg \lg W)$ follows from work of Even, Naor, Rao, & Schieber [9]. We shall present a series of results that give constant approximation algorithms for this problem, where the resulting algorithms are both simple to state, and simple to analyze.

We shall also consider the uncapacitated facility location problem. In this problem, there is a set of locations F at which we may build a facility (such as a warehouse), where the cost of building at location i is f_i, for each $i \in F$. There is a set D of client locations (such as stores) that require to be serviced by a facility, and if a client at location j is assigned to a facility at location i, a cost of c_{ij} is incurred. All of the data are assumed to be non-negative. The objective is to determine a set of locations at which to open facilities so as to minimize the total facility and assignment costs.

Building on results for the set covering problem (due to Johnson [19], Lovász [25], and Chvátal [7]), Hochbaum [15] showed that a simple greedy heuristic is an $O(\log n)$-approximation algorithm, where n denotes the total number of locations in the input. Lin & Vitter [24] gave an elegant filtering and rounding technique that yields an alternate $O(\log n)$-approximation algorithm for this problem. We shall focus on the metric case of this problem, in which distances between locations are given in some metric (and hence satisfy the triangle inequality), and the assignment costs c_{ij} are proportional to the distance between i and j, for each $i \in F$, $j \in D$. We shall present a series of results that give constant approximation algorithms for this problem, where, once again, the resulting algorithms are both simple to state, and (relatively) simple to analyze.

2   A simple scheduling problem

We shall present approximation algorithms for the problem of scheduling precedence-constrained jobs on a single machine so as to minimize the average weighted completion time, $1|prec| \sum w_j C_j$. Although we will primarily focus on this one scheduling model, the starting point for the work that we shall survey is an extremely simple, elegant result of Phillips, Stein, & Wein [29] for a related problem, in which the jobs are now independent (that is, there are no precedence constraints) but instead each job j has a specified release date r_j before which it may not begin processing, $j = 1, \ldots, n$; furthermore, they consider the unit-weight case, or in other words, $w_j = 1$, for each $j = 1, \ldots, n$. This problem is denoted $1|r_j| \sum C_j$ and was shown to be $\mathcal{NP}$-hard by Lenstra, Rinnooy Kan, & Brucker [22].

The algorithm of Phillips, Stein, & Wein [29] is based on a relaxation of the problem that can be solved in polynomial time. In this case, however, the relaxation is not a linear program, but instead one motivated in purely scheduling terms: rather than requiring that each job be processed without interruption, we allow *preemption*. That is, the processing of a job may be interrupted to process another (higher priority) job instead, and then the first job may be resumed without penalty. This problem, denoted $1|r_j, pmtn| \sum C_j$, can be solved (to optimality) by the following simple rule: schedule the jobs in time, and always process the job with the least remaining processing time (among those already released).

The approximation algorithm of Phillips, Stein, & Wein works as follows: solve the preemptive relaxation, and then schedule the jobs in the order in which they complete in the relaxed solution. It is remarkably straightforward to show that this is a 2-approximation algorithm. Suppose that the jobs happen to be indexed in the order in which they complete in the preemptive relaxation, and so are processed in the order $1, 2, \ldots, n$ in the heuristically computed non-preemptive schedule as well. If we consider the schedule produced by the approximation algorithm, then any idle time in the schedule ends at the release date of some job k (since that idle time is, in effect, caused by waiting for job k to be released). Consequently, for each job j, there is no idle time between $\max_{k=1,\ldots,j} r_k$ and the completion time of job j, C_j. This implies that

$$C_j \leq \max_{k=1,\ldots,j} r_k + \sum_{k=1}^{j} p_j.$$

Let $\overline{C}_j$ denote the completion time of job j in the optimal preemptive schedule; since each job k, $k = 1, \ldots, j$, has completed its processing in the optimal preemptive schedule by $\overline{C}_j$, it follows that

$$r_k \leq \overline{C}_k \leq \overline{C}_j, \quad \text{for each } k = 1, \ldots, j,$$

By the same reasoning, $\sum_{k=1}^{j} p_k \leq \overline{C}_j$. Hence, $C_j \leq 2\overline{C}_j$. Furthermore, the value of the schedule found, $\sum_{j=1}^{n} C_j$, is at most twice the preemptive optimum, and so is at most twice the value of the non-preemptive optimal schedule as well.

For $1|prec| \sum w_j C_j$, we shall rely on a number of linear programming relaxations, but the overall approach will be identical. We will solve the relaxation, and then use the relaxed solution to compute a (natural) ordering of the jobs that is feasible with respect to $\prec$; this is the schedule computed by the approximation algorithm. This is not the first scheduling problem for which this approach has been considered; for example, Munier & König [28] have given a very elegant approximation algorithm where the schedule (for a particular parallel machine scheduling problem with communication delays) is derived from an optimal solution to a linear programming relaxation.

We start by considering a very strong linear programming relaxation, the *non-preemptive time-indexed formulation*. In this formulation, which is due to Dyer & Wolsey [8], we use the variable x_{jt} to indicate whether job j completes processing at time t, $j = 1, \ldots, n$, $t = 1, \ldots, T$, where $T = \sum_{j=1}^{n} p_j$. Given these decision variables, it is easy to represent the objective function:

$$\text{Minimize} \sum_{j=1}^{n} w_j \sum_{t=1}^{T} t \cdot x_{jt}. \tag{1}$$

We can constrain the assignments of the decision variables as follows. Each job must complete at a unique point in time; hence,

$$\sum_{t=1}^{T} x_{jt} = 1, \quad j = 1, \ldots, n. \tag{2}$$

No job j can complete before p_j:

$$x_{jt} = 0, \quad \text{if } t < p_j. \tag{3}$$

The sum $\sum_{s=1}^{t} x_{js} = 1$ if and only if job j has been completed by time t; if $j \prec k$, we know that job j must complete at least p_k time units earlier than job k, and hence

$$\sum_{s=1}^{t} x_{js} \geq \sum_{s=1}^{t+p_k} x_{ks}, \quad \text{if } j \prec k, \ t = 1, \ldots, T - p_k. \tag{4}$$

Of course, the machine can process at most one job at each time t; job j is processed at time t if it completes at any time within the interval $[t, t + p_j - 1]$:

$$\sum_{j=1}^{n} \sum_{s=t}^{t+p_j-1} x_{js} \leq 1, \quad t = 1, \ldots, T. \tag{5}$$

If we wish to give an integer programming formulation of the problem, then we would require each variable to be either 0 or 1. We shall consider the linear programming relaxation, in which we require that $x_{jt} \geq 0$, $j = 1, \ldots, n$, $t = 1, \ldots, T$. For any feasible fractional solution x, we define $\overline{C}_j = \sum_{t=1}^{T} t \cdot x_{jt}$ to be the *fractional completion time* of job j, $j = 1, \ldots, n$. If x is an optimal solution to the linear relaxation, then $\sum_{j=1}^{n} w_j \overline{C}_j$ is a lower bound on the optimal value for the original problem.

For a given α, $0 \leq \alpha \leq 1$, and a job j, $j = 1, \ldots, n$, we focus on the earliest point in time that a cumulative α-fraction of job j has been slated to complete: let the *α-point* of job j be $t_j(\alpha) = \min\{t : \sum_{s=1}^{t} x_{js} \geq \alpha\}$. The notion of an α-point was also introduced in the work of Phillips, Stein, & Wein [29], in a slightly different context. Hall, Shmoys, & Wein [14] proposed the following algorithm for $1|prec| \sum w_j C_j$: schedule the jobs in non-decreasing order of their α-points. It is easy to see that the constraints (4) ensure that the schedule found satisfies the precedence constraints.

The α-*point algorithm* of Hall, Shmoys, & Wein can be analyzed as follows. Suppose that the jobs happen to be indexed in non-decreasing α-point order. Hence, each job j completes at time

$$C_j = \sum_{k=1}^{j} p_k. \tag{6}$$

For each job k, $k = 1, \ldots, j$, an α fraction of each job k is done by time $t_j(\alpha)$, and hence

$$\alpha \sum_{k=1}^{j} p_k \leq t_j(\alpha). \tag{7}$$

Consider the fractional completion time $\overline{C}_j$; one can view the values x_{jt} as providing a weighted average of the corresponding values t. Since less than a $1 - \alpha$ fraction of the weight can be placed on values more than $1/(1 - \alpha)$ times the average, we see that

$$t_j(\alpha) \leq \overline{C}_j/(1 - \alpha). \tag{8}$$

By combining (6)–(8), we see that each job j completes at time

$$C_j \leq \overline{C}_j/(\alpha(1 - \alpha)).$$

Consequently, we see that the value of the solution found, $\sum_{j=1}^{n} w_j C_j$, is within a factor of $1/(\alpha - \alpha^2)$ of $\sum_{j=1}^{n} w_j \overline{C}_j$, which is a lower bound on the optimal value. If we set $\alpha = 1/2$ (to minimize $1/(\alpha - \alpha^2)$), we see that we have obtained a solution of value within a factor of 4 of the optimum.

But is setting $\alpha = 1/2$ the best thing to do? Goemans [10] observed that rather than choosing α once, to optimize the performance guarantee, it makes more sense to consider, for each input, which choice of α would deliver the best schedule for *that particular input*. (Chekuri, Motwani, Natarajan, & Stein [3] independently suggested an analogous improvement to the algorithm of Phillips, Stein, & Wein.) The performance of this best-α algorithm can be analyzed by considering the following randomized algorithm instead: set $\alpha = a$ by choosing at random within the interval (0,1) according to the probability density function $f(a) = 2a$. The same analysis given above implies that we can bound

$$E[C_j] \leq \int_0^1 (t_j(a)/a)f(a)da = 2 \int_0^1 t_j(a)da.$$

If we interpret this integral as the area under the curve defined by the function $t_j(a)$ as a ranges from 0 to 1, then it is easy to see that this integral is precisely $\overline{C}_j$. Thus, the randomized algorithm produces a solution that has expected value at most *twice* the optimal value. Furthermore, the algorithm that finds the value of α for which the α-point algorithm delivers the best solution, the *best-α algorithm*, is a deterministic algorithm guaranteed to find a solution with objective function value at most twice the optimal value.

Of course, none of these algorithms are efficient; that is, it is not known how to implement them to run in polynomial time, due to the size of the linear programs that must be solved. Since the size of the linear program can be bounded by a polynomial in n and $T = \sum_j p_j$, the α-point algorithm can be shown to run in pseudo-polynomial time. It is often the case that a pseudo-polynomial algorithm for a problem can be adapted to run in polynomial time while losing an additional $1 + \epsilon$ factor in accuracy, basically by using only a polynomial number of bits of accuracy in the input. However, in this case it is not clear how to use to these well-known techniques.

Instead, Hall, Shmoys, & Wein [14] proposed using a different, more compact, linear programming relaxation, called an *interval-indexed formulation*. (This type of formulation was subsequently used in another context in the journal version of these results [13].) The key idea behind these constructions is that the time horizon is subdivided into the intervals $[1, 1]$, $(1, 1 + \epsilon]$, $(1 + \epsilon, (1 + \epsilon)^2]$, $((1 + \epsilon)^2, (1 + \epsilon)^3], \ldots$, where ϵ is an arbitrarily small positive constant; the linear program only specifies the interval in which a job is completed. Since all completion times within an interval are within a $(1 + \epsilon)$ factor of each other, the relative scheduling within an interval will be of little consequence.

Given this basic idea, it is extremely straightforward to complete all of the details of this polynomial-sized formulation. The linear programming relaxation relies on the variables $x_{j\ell}$, which indicate whether job j completes within the ℓth interval. There are assignment constraints completely analogous to (2). The precedence constraints are enforced only to the extent that if $j \prec k$, then the interval in which j finishes is no later than the interval in which k finishes. To capture the load constraint, we merely require that the total length of jobs assigned to complete in the interval $((1 + \epsilon)^{\ell-1}, (1 + \epsilon)^\ell]$ is at most $(1 + \epsilon)^\ell$. The analogue of the α-point algorithm is as follows: for each job, compute its α-interval, and schedule the jobs in order of non-decreasing α-intervals, where the jobs assigned to the same interval are scheduled in any order that is consistent with the precedence relation. Thus, Hall, Shmoys, & Wein obtained, for any fixed $\epsilon > 0$, a $4 + \epsilon$-approximation algorithm, and the best-α-point algorithm of Goemans can be adapted to yield a $2 + \epsilon$-approximation algorithm.

As it turns out, it is even easier to obtain a 2-approximation algorithm for this problem by using other compact linear programming relaxations. Schulz [35] (and subsequently in its journal version [13]) showed how to improve the earlier work of Hall, Shmoys, & Wein by using a relaxation due to Wolsey [41] and Queyranne [31]. In this formulation, there is a variable C_j for each job j in $N = \{1, \ldots, n\}$:

$$\text{Minimize} \sum_{j=1}^n w_j C_j \tag{9}$$

subject to

$$\sum_{j \in S} p_j C_j \geq \sum_{(j,k) \in S \times S} p_j p_k, \qquad \text{for each } S \subseteq N, \tag{10}$$

$$C_k \geq C_j + p_k, \qquad \text{if } j \prec k. \tag{11}$$

If the jobs are independent, and hence there are neither precedence constraints nor constraints in (11), then Wolsey [41] and Queyranne [31] independently showed that this linear program provides an exact characterization of the problem $1||\sum w_j C_j$: extreme points of this linear program correspond to schedules. Of course, in the case in which there are precedence constraints, the situation is quite different, since otherwise $\mathcal{P}$ would be equal to $\mathcal{NP}$.

The most natural approximation algorithm for $1|prec|\sum w_j C_j$ based on this linear relaxation is as follows: solve the relaxation to obtain a solution $\overline{C}_j$, $j = 1, \ldots, n$, and schedule the jobs so that their LP values are in non-decreasing order. The analysis of this algorithm is also remarkably simple. Suppose that the jobs happen to be indexed so that $\overline{C}_1 \leq \cdots \leq \overline{C}_n$, and so they are scheduled by the algorithm in their index order as well. Once again, job j completes at time $C_j = \sum_{k=1}^{j} p_k$. If we consider the constraint (10) when $S = \{1, \ldots, j\}$, then we see that

$$\sum_{k=1}^{j} p_k \overline{C}_k \geq \sum_{(k,k')\in S \times S} p_k p_{k'} \geq (1/2)(\sum_{k=1}^{j} p_k)^2.$$

However, $\overline{C}_j(\sum_{k=1}^{j} p_k) \geq \sum_{k=1}^{j} p_k \overline{C}_k$. Hence $\overline{C}_j \geq (\sum_{k=1}^{j} p_k)/2$, or equivalently, $C_j \leq 2\overline{C}_j$. This proves that the value of the solution found is within a factor of 2 of optimal. However, it is not at all clear that this linear programming relaxation is sufficiently more compact than the time-indexed one, since it contains an *exponential* number of constraints. However, one can solve this linear program in polynomial time with the ellipsoid algorithm, since it is easy to devise a polynomial-time algorithm that determines whether a given fractional solution is feasible, or if not, returns a violated constraint (see Queyranne [31]). Hence, we have a 2-approximation algorithm.

Potts [30] has proposed yet another linear programming relaxation of the problem $1|prec|\sum w_j C_j$, which is called the *linear ordering formulation*. In this formulation, there are variables δ_{ij} that indicate whether or not job i is processed before job j:

$$\text{Minimize } \sum_{j=1}^{n} w_j C_j$$

subject to

$$p_j + \sum_{i=1}^{n} p_i \delta_{ij} = C_j, \quad j = 1, \ldots, n;$$
$$\delta_{ij} + \delta_{ji} = 1, \quad i, j = 1, \ldots, n, \ i < j;$$
$$\delta_{ij} + \delta_{jk} + \delta_{ki} \leq 2, \quad i, j, k = 1, \ldots, n, \ i < j < k \text{ or } i > j > k;$$
$$\delta_{ij} = 1, \quad i, j = 1, \ldots, n, \ i \prec j;$$
$$\delta_{ij} \geq 0, \quad i, j = 1, \ldots, n, \ i \neq j.$$

Schulz [35] has observed that for any feasible solution to this linear program, the C_j values are feasible for the linear program (9)–(11). Hence, if we solve the linear ordering formulation to obtain values $\overline{C}_j$, and then schedule the jobs so that these values are in non-decreasing order, then we obtain a more efficient

2-approximation algorithm (since any polynomial-time linear programming algorithm can be used to solve this LP with n^2 variables and $O(n^3)$ constraints).

Chudak & Hochbaum [5] proposed a somewhat weaker linear programming relaxation, which also uses the variables δ_{ij}. In this relaxation, the constraints that enforce the transitivity of the ordering relaxation, $\delta_{ij} + \delta_{jk} + \delta_{ki} \leq 2$, are instead replaced with the constraints that $\delta_{ki} \leq \delta_{kj}$, whenever $i \prec j$, and k is different from both jobs i and j. Once again, a straightforward calculation shows that for any feasible solution to this weaker linear program, the C_j values are feasible for the constraints (10) and (11). Consequently, one also obtains a 2-approximation algorithm by first solving this weaker linear program, and then using the resulting C_j values to order the jobs. The advantage of using this formulation is as follows: Chudak & Hochbaum also observed that a result of Hochbaum, Meggido, Naor, & Tamir [17] can be applied to show that there always exists an optimal solution to this linear program that is *half-integral*, i.e., each variable δ_{ij} is either 0,1/2, or 1; furthermore, an optimal half-integral solution can be computed by a maximum flow computation. Thus, this approach yields a 2-approximation algorithm that does not require the solution of a linear program, but rather only a single maximum flow computation.

Chekuri & Motwani [2] and Margot, Queyranne, & Wang [27] independently devised another, more combinatorial 2-approximation algorithm for the problem $1|prec|\sum w_j C_j$. We shall say that a subset S of jobs is an *initial set* of the precedence relation $\prec$ if, for each job $k \in S$, each of its predecessors is also in S, or more formally,

$$(k \in S \text{ and } j \prec k) \Rightarrow j \in S.$$

For each subset of jobs $S \subseteq N$, let $\rho(S) = \sum_{j \in S} p_j / \sum_{j \in S} w_j$.

Suppose that we minimize $\rho(S)$ over all initial subsets to obtain a subset S^*. Chekuri & Motwani and Margot, Queyranne, & Wang proved a remarkable fact: if $S^* = N$, then *any* ordering of the jobs that is consistent with $\prec$ has objective function value within a factor of 2 of the optimum. The proof of this fact is amazingly simple. In each feasible schedule, each job j completes by time $\sum_{k \in N} p_k$, and so the cost of any solution is at most $(\sum_{k \in N} p_k)(\sum_{k \in N} w_k)$. So we need only show that the optimal value is at least $(\sum_{k \in N} p_k)(\sum_{k \in N} w_k)/2$. Suppose that the jobs happen to be indexed so that job j is the jth job to be scheduled in an optimal schedule. Then each set $\{1, \ldots, j\}$ is an initial set, and hence the completion time of job j,

$$C_j = \sum_{k=1}^{j} p_k \geq \rho(N) \sum_{k=1}^{j} w_k.$$

Consequently, we know that

$$\sum_{j=1}^{n} w_j C_j \geq \rho(N) \sum_{j=1}^{n} \sum_{k=1}^{j} w_j w_k \geq \rho(N)(\sum_{j=1}^{n} w_j)^2/2.$$

Recalling that $\rho(N) = \sum_{j=1}^{n} p_j / \sum_{j=1}^{n} w_j$, we see that we have obtained the desired lower bound on the optimal value.

Of course, there is no reason to believe that N is the initial set S for which $\rho(S)$ is minimized. Fortunately, if this is not the case, then we can rely on the following decomposition result of Sidney [37]: if S^* is the initial set S for which $\rho(S)$ is minimized, then there exists an optimal solution in which the jobs of S^* precede the jobs of $N - S^*$. This suggests the following recursive 2-approximation algorithm: find the set S^*, and schedule it first in any order consistent with the precedence relation $\prec$, and then recursively apply the algorithm to $N - S^*$, and concatenate the two schedules found. It is not hard to show that the initial set S^* can be found via a minimum cut (or equivalently, a maximum flow) computation.

For each of the results above, we have presented an algorithm and then showed that it delivers a solution whose objective function value is within some constant factor of the optimal value of a linear programming relaxation of the problem. Such a result not only shows that we have found a good algorithm, but also implies a guarantee for the quality of the lower bound provided by that linear program. For each of the linear programs concerned, one might ask whether these particular algorithms can be improved; that is, might it be possible to round the optimal fractional solutions in a more effective manner? Unfortunately, the answer to each of these questions is no. For the time-indexed formulation, Schulz & Skutella [34] have given instances for which the ratio between the integer and fractional optima is arbitrarily close to 2. For the linear ordering formulation, Chekuri & Motwani [2] have given a surprising construction based on expander graphs for which the ratio of the integer to fractional optimal values asymptotically approaches 2. Each of these results implies the analogous result for the linear program (9)–(11), but for this relaxation it is also relatively simple to construct examples directly. Of course, there might still be other relaxations that provide stronger lower bounds, and this is an extremely interesting direction for further research.

3   The uncapacitated facility location problem

The uncapacitated facility location problem is one of the most well-studied problems in the Operations Research literature, dating back to the work of Balinski [1], Kuehn & Hamburger [20], Manne [26], and Stollsteimer [38, 39] in the early 60's. We shall focus on one important special case of this problem, where the locations are embedded in some metric space, and the assignment costs c_{ij} are proportional to the distances between locations; we shall call this the *metric uncapacitated facility location problem*.

Although there is little work that has specifically focused on the metric case of this location problem, for many others, such as the k-center problem (see, e.g., [18]) and the k-median problem (see, e.g., [23]) this assumption is prevalent. In fact, the algorithms of Lin & Vitter [23] contained many of the seeds of the work that we shall present for the metric uncapacitated facility location problem.

Once again, all of the algorithms that we shall discuss will be based on rounding an optimal solution to a linear programming relaxation of the problem. For this problem, the most natural relaxation is as follows. There are two types of decision variables x_{ij} and y_i, for each $i \in F$, $j \in D$, where each variable y_i, $i \in F$, indicates whether or not a facility is built at location i, and each variable x_{ij} indicates whether or not the client at location j is assigned to a facility at location i, for each $i \in F$, $j \in D$:

$$\text{Minimize } \sum_{i \in F} f_i y_i + \sum_{i \in F} \sum_{j \in D} c_{ij} x_{ij} \tag{12}$$

subject to

$$\sum_{i \in F} x_{ij} = 1, \qquad \text{for each } j \in D, \tag{13}$$

$$x_{ij} \leq y_i, \qquad \text{for each } i \in F, \, j \in D, \tag{14}$$

$$x_{ij} \geq 0, \qquad \text{for each } i \in F, \, j \in D. \tag{15}$$

Shmoys, Tardos, & Aardal [36] gave a simple algorithm to round an optimal solution to this linear program to an integer solution of cost at most $3/(1-e^3) \approx 3.16$ times as much. The algorithm relies on the filtering technique of Lin & Vitter [24]. We can interpret each fractional solution (x, y) as the following bipartite graph $G(x, y) = (F, D, E)$: the two sets of nodes are F and D, and there is an edge $(i, j) \in E$ exactly when $x_{ij} > 0$.

First, we apply an α-filtering algorithm to convert the optimal fractional solution to a new one, $(\bar{x}, \bar{y})$, in which the cost c_{ij} associated with each edge in $G(\bar{x}, \bar{y})$ is relatively cheap. As in the algorithm based on the time-indexed formulation for the scheduling problem, we first define the notion of an α-point, $c_j(\alpha)$, for each location $j \in D$. Focus on a location $j \in D$, and let π be a permutation such that $c_{\pi(1)j} \leq c_{\pi(2)j} \leq \cdots \leq c_{\pi(n)j}$. We then set $c_j(\alpha) = c_{\pi(i^*)j}$, where $i^* = \min\{i' : \sum_{i=1}^{i'} x_{\pi(i)j} \geq \alpha\}$. To construct $(\bar{x}, \bar{y})$, for each $(i, j) \in E(x, y)$ for which $c_{ij} > c_j(\alpha)$ we set $\bar{x}_{ij} = 0$, and then renormalize by setting each remaining $\bar{x}_{ij}$ equal to x_{ij}/α_j, where $\alpha_j = \sum_{(i,j) \in E: \, c_{ij} \leq c_j(\alpha)} x_{ij}$. We also renormalize $\bar{y}_i = y_i/\alpha$. It is easy to check that $(\bar{x}, \bar{y})$ is a feasible solution to the linear program (12)–(15) with the further property that $\bar{x}_{ij} > 0 \Rightarrow c_{ij} \leq c_j(\alpha)$. Motivated by this, given values g_j, $j \in D$, we shall call a solution g-close if $\bar{x}_{ij} > 0 \Rightarrow c_{ij} \leq g_j$.

The central element of the rounding algorithm of Shmoys, Tardos, & Aardal is a polynomial-time algorithm that, given a g-close feasible solution $(\bar{x}, \bar{y})$ to (12)–(15), finds a $3g$-close integer solution $(\hat{x}, \hat{y})$ such that

$$\sum_{i \in F} f_i \hat{y}_i \leq \sum_{i \in F} f_i \bar{y}_i.$$

The algorithm works as follows. It partitions the graph $G(\bar{x}, \bar{y}) = (F, D, E)$ into clusters, and then, for each cluster, opens one facility that must serve all clients in it. The clusters are constructed iteratively as follows. Among all clients that have not already been assigned to a cluster, let j' be the client j for which g_j is smallest. This cluster consists of j', all neighbors of j' in $G(\bar{x}, \bar{y})$, and all of their neighbors as well (that is, all nodes j such that there exists some i for which

(i, j) and (i, j') are both in E. Within this cluster, we open the cheapest facility i' and use it to serve all clients within this cluster.

We next show that this rounding algorithm has the two claimed properties. Each client j in the cluster is assigned to a facility i' for which there is a path in $G(\bar{x}, \bar{y})$ consisting of an edge connecting i' and j' (of cost at most $g_{j'}$), an edge connecting j' and some node i (of cost at most $g_{j'}$), and an edge connecting i and j (of cost at most g_j). Hence, by the triangle inequality, the cost of assigning j to i' is at most $2g_{j'} + g_j$. Since j was chosen as the remaining client with minimum g-value, it follows that $g_{j'} \leq g_j$, and so the cost of assigning j to i' is at most $3g_j$. In other words, the integer solution found is $3g$-close.

Consider the first cluster formed, and let j' be the node with minimum g-value used in forming it. We know that $\sum_{i:(i,j')\in E} \bar{x}_{ij'} = 1$. Since the minimum of a set of values is never more than a weighted average of them, the cost of the facility selected

$$f_{i'} \leq \sum_{i:(i,j')\in E} \bar{x}_{ij'} f_i \leq \sum_{i:(i,j')\in E} \bar{y}_i f_i,$$

where the last inequality follows from constraint (14). Observe that, throughout the execution of the algorithm, each location $j \in D$ that has not yet been assigned to some cluster, has the property that each of its neighbors i must also remain unassigned. Hence, for each cluster, the cost of its open facility is at most the cost that the fractional solution assigned to nodes in F within that cluster. Hence, in total,

$$\sum_{i\in F} f_i \hat{y}_i \leq \sum_{i\in F} f_i \bar{y}_i.$$

Thus, we have argued that the rounding algorithm of Shmoys, Tardos, & Aardal has the two key properties claimed above.

Suppose that we apply this rounding theorem to an α-filtered solution. What can we prove about the cost of the resulting integer solution? By the two properties proved above, we know that the cost of the solution is at most

$$\sum_{i\in F} f_i \hat{y}_i + \sum_{i\in F}\sum_{j\in D} c_{ij} \hat{x}_{ij} \leq \sum_{i\in F} f_i \bar{y}_i + \sum_{j\in D} 3c_j(\alpha) = \sum_{i\in F} f_i y_i/\alpha + 3\sum_{j\in D} c_j(\alpha).$$

However, exactly analogous to (8), we again know that at most a $(1-\alpha)$ fraction of the values in a weighted average can exceed $1/(1-\alpha)$ times the average, and hence

$$c_j(\alpha) \leq (\sum_{i\in D} c_{ij} x_{ij})/(1-\alpha).$$

Plugging this bound into the previous inequality, we see that the total cost of the solution found is at most

$$\max\{\frac{1}{\alpha}, \frac{3}{1-\alpha}\}(\sum_{i\in F} f_i y_i + \sum_{i\in F}\sum_{j\in D} c_{ij} x_{ij}).$$

If we set $\alpha = 1/4$, then we see that the total cost of the solution found is at most 4 times the cost of (x, y), and so by rounding an optimal solution to the linear relaxation, we obtain a 4-approximation algorithm.

Once again, we may apply the idea of Goemans [10]; it is foolish to set α once, rather than choosing the best α for each input. Once again, we will analyze this best-α algorithm by analyzing a randomized algorithm instead. Let $0 < \beta < 1$ be a parameter to be fixed later. We shall set $\alpha = a$, where a is selected uniformly at random within the interval $[\beta, 1]$. Once again, we shall rely on the fact that

$$\int_0^1 c_j(a)da = \sum_{i=1}^n c_{ij}x_{ij}.$$

The expected cost of the solution found can be upper bounded by

$$E[\frac{1}{a}\sum_{i \in F} f_i y_i + 3 \sum_{j \in D} c_j(a)] = E[\frac{1}{a}] \sum_{i \in F} f_i y_i + 3 \sum_{j \in D} E[c_j(a)]$$

$$= (\int_\beta^1 \frac{1}{1-\beta}\frac{1}{a}da) \sum_{i \in F} f_i y_i + 3 \sum_{j \in D}(\int_\beta^1 \frac{1}{1-\beta}c_j(a)da)$$

$$\leq \frac{\ln(1/\beta)}{1-\beta} \sum_{i \in F} f_i y_i + \frac{3}{1-\beta} \sum_{j \in D} \int_0^1 c_j(a)da$$

$$= \frac{\ln(1/\beta)}{1-\beta} \sum_{i \in F} f_i y_i + \frac{3}{1-\beta} \sum_{j \in D}\sum_{i \in F} c_{ij}x_{ij}.$$

If we set $\beta = 1/e^3$, then we have obtained the claimed $\frac{3}{1-e^3}$-approximation algorithm.

Guha & Khuller [12] proposed the following improvement to the algorithm of Shmoys, Tardos, & Aardal. A natural way in which to compute a better solution is to perform a post-processing phase in which one iteratively checks if an additional facility can be opened to reduce the overall cost, and if so, greedily opens the facility that most reduces the total cost. Furthermore, Guha & Khuller also proposed the following strengthening of the linear programming relaxation. If one knew the cost ϕ incurred to build facilities in the optimal solution, one could add the constraint that $\sum_{i \in F} f_i y_i \leq \phi$. Since we don't know this value, we can instead guess this value by setting ϕ equal to $(1 + \epsilon)^k$, for each $k = 1, \ldots, \log_{1+\epsilon} \sum_{i \in F} f_i$, where ϵ is an arbitrarily small positive constant. There are only a polynomial number of settings for ϕ that must be considered, and so, in effect, we may assume that we know the correct ϕ to an arbitrary number of digits of accuracy. By adding the post-processing phase to the result of applying the rounding algorithm to the strengthened relaxation, Guha & Khuller obtain a 2.408-approximation algorithm. Guha & Khuller [12] and Sviridenko [40] independently showed that this problem is MAXSNP-hard, and hence there exists some constant $\rho > 1$ for which no ρ-approximation algorithm exists, unless $\mathcal{P} = \mathcal{NP}$. Guha & Khuller also showed a much stronger result, that no approximation algorithm can have performance guarantee better than 1.463 (unless $\mathcal{NP} \subseteq DTIME(n^{O(\log \log n)})$).

Chudak & Shmoys, independently, obtained a more modest improvement, a 3-approximation algorithm, which relies only on the original linear programming relaxation. The first essential idea in their improvement was the observation that the filtering step is, in some sense, completely unnecessary for the performance of the algorithm. This was based on a simple property of the optimal solution to the linear programming relaxation. Consider the dual to the linear program (12)–(15):

$$\text{Maximize} \sum_{j \in D} v_j \tag{16}$$

subject to

$$\sum_{j \in D} w_{ij} \leq f_i, \qquad \text{for each } i \in F,$$

$$v_j - w_{ij} \leq c_{ij}, \qquad \text{for each } i \in F, \ j \in D,$$

$$w_{ij} \geq 0 \qquad \text{for each } i \in F, \ j \in D.$$

This dual can be motivated in the following way. Suppose that we wish to obtain a lower bound for our input to the uncapacitated facility location problem. If we reset all fixed costs f_i to 0, and solve this input, then clearly we get a (horrible) lower bound: each client $j \in D$ gets assigned to its closest facility at a cost of $\min_{i \in F} c_{ij}$. Now suppose we do something a bit less extreme. Each location $i \in F$ decides on a given cost-sharing of its fixed cost f_i. Each location $j \in D$ is allocated a share w_{ij} of the fixed cost; if j is assigned to an open facility at i, then it must pay an additional fee of w_{ij} (for a total of $c_{ij} + w_{ij}$), but the explicit fixed cost of i is once again reduced to 0. Of course, we insist that each $w_{ij} \geq 0$, and $\sum_{j \in D} w_{ij} \leq f_i$ for each $i \in F$. But this is still an easy input to solve: each $j \in D$ incurs a cost $v_j = \min_{i \in F}(c_{ij} + w_{ij})$, and the lower bound is $\sum_{j \in D} v_j$. Of course, we want to allocate the shares so as to maximize this lower bound, and this maximization problem is precisely the LP dual.

Consider a pair of primal and dual optimal solutions: (x, y) and (v, w). Complementary slackness implies that if $x_{ij} > 0$, then the corresponding dual constraint is satisfied with equality. That is, $v_j - w_{ij} = c_{ij}$, and since $w_{ij} \geq 0$, we see that $c_{ij} \leq v_j$; in other words, (x, y) is already v-close. Hence, if we apply the rounding algorithm of Shmoys, Tardos, & Aardal (without filtering first, and so $g_j = v_j$), we find a solution of cost at most

$$\sum_{i \in F} f_i y_i + \sum_{j \in D} 3v_j = \sum_{i \in F} f_i y_i + 3\left(\sum_{i \in F} f_i y_i + \sum_{i \in F}\sum_{j \in D} c_{ij} x_{ij}\right) \leq 4\left(\sum_{i \in F} f_i y_i + \sum_{i \in F}\sum_{j \in D} c_{ij} x_{ij}\right),$$

where the first equality follows from the fact that the optimal solutions to the primal and the dual linear programs have equal objective function values.

The second key idea in the improvement of Chudak & Shmoys was the use of randomized rounding in the facility selection step. Randomized rounding is an elegant technique introduced by Raghavan & Thompson [32], in which a feasible solution to a linear programming relaxation of a 0–1 integer program is rounded to an integer solution by interpreting the fractions as probabilities, and setting each variable to 1 with the corresponding probability. Sviridenko [40] proposed a simple randomized rounding approximation algorithm for the special case of the metric uncapacitated facility location problem in which each $c_{ij} \in \{1, 2\}$. In the deterministic algorithm presented above, the cheapest facility in each cluster was opened. Instead, if the cluster is "centered" at j', one can open facility i with probability $x_{ij'}$. This does not really change the previous analysis, since the expected cost of the facilities selected is at most $\sum_{i \in F} f_i y_i$, and the bound on the assignment costs was independent of the choice of the facility opened in each cluster.

The final idea used to obtain the improved performance guarantee is as follows: rather than select the next center by finding the remaining client for which v_j is minimum (since $g_j = v_j$ in the version without filtering), select the client for which $v_j + \sum_{i \in F} c_{ij} x_{ij}$ is minimum. This enters into the analysis in the following way. For each client j in the cluster "centered" at j', its assignment cost is bounded by the cost of an edge (i, j) (of cost at most v_j), an edge (i, j') (of cost at most $v_{j'}$), and the edge (i', j'). The last of these costs is a random variable, and so we can focus on its expected value. Since j' chooses to open each facility i with probability $x_{ij'}$, the expected cost of the edge (i', j') is exactly $\sum_{i \in F} c_{ij'} x_{ij'}$. Thus, the expected cost of assigning j to i' is at most $v_j + v_{j'} + \sum_{i \in F} c_{ij'} x_{ij'}$. By our modified selection rule, this expectation is at most $2v_j + \sum_{i \in F} c_{ij} x_{ij}$, and hence the expected total cost of the solution is at most

$$\sum_{j \in D} 2v_j + \sum_{j \in D} \sum_{i \in F} c_{ij} x_{ij} + \sum_{i \in F} f_i y_i,$$

which is exactly equal to three times the optimal value of the linear programming relaxation.

The analogous deterministic algorithm is quite natural. Before, we merely chose the cheapest facility in each cluster. However, by choosing a facility, we also affect the assignment cost of each client in that cluster. Thus, if choose the facility that minimizes the *total cost for that cluster*, then we achieve a deterministic 3-approximation algorithm.

However, this is not the best possible analysis of this randomized algorithm. Subsequently, Chudak [4] and Chudak & Shmoys [6] have improved this bound to show that (essentially) this randomized algorithm leads to a $(1 + 2/e)$-approximation algorithm. We shall modify the algorithm in the following way. For each location $i \in F$, there is some probability p_i with which it has been opened by this algorithm. (For most locations, it is equal to some value $x_{ij'}$ when facility location i belongs to a cluster "centered" at j', but some locations i might not belong to any cluster.) In the modified algorithm, we also have independent events that open each facility i with probability $y_i - p_i$. In fact, we can simplify some of this discussion by making the following further assumption about the optimal solution (x, y) to the linear program (12)–(15): for each $x_{ij} > 0$, it follows that $x_{ij} = y_i$. We shall say that such a solution is *complete*. This assumption can be made without loss of generality, since it is not hard to show that for any input, there is an equivalent input for which the optimal fractional solution is complete.

For the algorithms above, we have indicated that each client is assigned to the facility that has been opened in its cluster. In fact, there is no need to make this assumption about the assignments, since we may simply assign each client to its cheapest open facility. Given this, the key insight to the improved analysis is as follows. Consider some client j (which is not the center of its cluster). We have shown that its assignment cost is at most $3v_j$ (for the 4-approximation algorithm, and a somewhat better bound for the 3-approximation algorithm). However, the randomized algorithm might very well open one of j's neighbors in $G(x, y)$. In that case, clearly we can obtain a much better bound on the assignment cost incurred for client j. In fact, one can show that the probability that a facility has been opened at least one of j's neighbors is at least $(1 - 1/e)$, and this is the basic insight that leads to the improved analysis.

Although the complete analysis of this algorithm is beyond the scope of this survey, we will outline its main ideas. The improvement in the bound is solely due to the fact that we can bound the expected assignment cost for each client j by $\sum_{i \in F} c_{ij} x_{ij} + (2/e)v_j$. In fact, we will only sketch the proof that this expectation is at most $\sum_{i \in F} c_{ij} x_{ij} + (3/e)v_j$, and will use as a starting point, the original clustering algorithm in which the next client selected is the one for which v_j is smallest (rather than the modified one in which selection was based on $v_j + \sum_{i \in F} c_{ij} x_{ij}$).

Suppose that the neighbors of client j in $G(x, y)$ happen to be nodes $1, \ldots, d$, where $c_{1j} \leq \cdots \leq c_{dj}$. Thus, $\sum_{i=1}^{d} x_{ij} = \sum_{i=1}^{d} y_i = 1$. We can bound the expected assignment cost for j, by considering nodes $i = 1, \ldots, d$ in turn, assigning j to the first of these that has been opened, and if none of these facilities have been opened, then assigning j to the "back-up" facility i' that has surely been opened in its cluster. If opening neighboring facilities $i = 1, \ldots, d$ were independent events, then a simple upper bound on the expected assignment cost for j is

$$y_1 c_{1j} + (1 - y_1) y_2 c_{2j} + \cdots + (1 - y_1) \cdots (1 - y_{d-1}) y_d c_{dj} + (1 - y_1) \cdots (1 - y_d) 3 v_j,$$

which is clearly at most $\sum_{i=1}^{d} c_{ij} y_i + 3 v_j \prod_{i=1}^{d} (1 - y_i)$. The Taylor series expansion of e^{-r} implies that $1 - r \leq e^{-r}$. Using this fact, and the assumption that the optimal LP solution (x, y) is complete, we see that the expected assignment cost for j is at most $\sum_{i \in F} c_{ij} x_{ij} + (3/e)v_j$.

However, opening the neighboring facilities $i = 1, \ldots, d$ are not independent events: for instance, if two of these neighbors are in the same cluster, then only one of them can be opened. The next question is: can the conditioning between these events be harmful? Fortunately, the answer is no, and it is fairly intuitive to see why this is the case. If it happens that none of the first k neighbors of j have not been opened, this only makes it more likely that the next cheapest facility is, in fact, open. A precise analysis of this situation can be given, and so one can prove that the expected assignment cost for j is at most $\sum_{i \in F} c_{ij} x_{ij} + (3/e)v_j$ (without relying on unsupportable assuptions).

These randomized approximation algorithms can each be derandomized, by a straightforward application of the method of conditional probabilities. Thus, if we return to the selection rule in which the next cluster is "centered" at the remaining client j for which $v_j + \sum_{i \in F} c_{ij} x_{ij}$ is minimized, then this derandomization leads to a $(1 + 2/e)$-approximation algorithm.

For the uncapacitated facility location problem, the natural questions for further research are even more tantalizing than for the scheduling problem discussed in the previous section. It is not known that the analysis of the algorithm of Chudak & Shmoys is tight (and in fact, we suspect that it is not tight). Guha & Khuller [12] have given an input for which the ratio between the optimal integer and fractional optima is at least 1.463, but this still leaves some room between that and the upper bound of $1 + 2/e \approx 1.736$ implied by the last algorithm. Furthermore, there are well-known ways to construct stronger linear programming relaxations for this problem, and it would be very interesting to use them to prove stronger performance guarantees.

References

1. M. L. Balinksi. On finding integer solutions to linear programs. In *Proceedings of the IBM Scientific Computing Symposium on Combinatorial Problems*, pages 225–248. IBM, 1966.
2. C. Chekuri and R. Motwani. Precedence constrained scheduling to minimize weighted completion time on a single machine. Unpublished manuscript, 1997.
3. C. Chekuri, R. Motwani, B. Natarajan, and C. Stein. Approximation techniques for average completion time scheduling. *Proceedings of the Eighth Annual ACM-SIAM Symposium on Discrete Algorithms*, pages 609–618, 1997.
4. F. A. Chudak. Improved approximation algorithms for uncapacitated facility location. In: *Proceedings of the 6th Integer Programming and Combinatorial Optimization Conference (IPCO)*, 1998, to appear.
5. F. A. Chudak and D. S. Hochbaum. A half-integral linear programming relaxation for scheduling precedence-constrained jobs on a single machine. Unpublished manuscript, 1997.
6. F. A. Chudak and D. B Shmoys. Improved approximation algorithms for the uncapacitated facility location problem. Unpublished manuscript, 1997.
7. V. Chvátal. A greedy heuristic for the set covering problem. *Math. Oper. Res.*, 4:233–235, 1979.
8. M. E. Dyer and L. A. Wolsey. Formulating the single machine sequencing problem with release dates as a mixed integer program. *Discrete Appl. Math.*, 26:255–270, 1990.
9. G. Even, J. Naor, S. Rao, and B. Schieber. Divide-and-conquer approximation algorithms via spreading metrics. In *Proceedings of the 36th Annual IEEE Symposium on Foundations of Computer Science*, pages 62–71, 1995.
10. M. X. Goemans. Personal communication, June, 1996.
11. R. L. Graham, E. L. Lawler, J. K. Lenstra, and A. H. G. Rinnooy Kan. Optimization and approximation in deterministic sequencing and scheduling: a survey. *Ann. Discrete Math.*, 5:287–326, 1979.
12. S. Guha and S. Khuller. Greedy strikes back: Improved facility location algorithms. In *Proceedings of the 9th Annual ACM-SIAM Symposium on Discrete Algorithms*, pages 649–657, 1998.

13. L. A. Hall, A. S. Schulz, D. B. Shmoys, and J. Wein. Scheduling to minimize the average completion time: on-line and off-line approximation algorithms. *Math. Oper. Res.*, 22:513–544, 1997.

14. L. A. Hall, D. B. Shmoys, and J. Wein. Scheduling to minimize the average completion time: on-line and off-line algorithms. In *Proceedings of the 7th Annual ACM-SIAM Symposium on Discrete Algorithms*, pages 142–151, 1996.

15. D. S. Hochbaum. Heuristics for the fixed cost median problem. *Math. Programming*, 22:148–162, 1982.

16. D. S. Hochbaum, editor. *Approximation algorithms for NP-hard problems*, Boston, MA, 1997. PWS.

17. D. S. Hochbaum, N. Megiddo, J. Naor, and A. Tamir. Tight bounds and 2-approximation algorithms for integer programs with two variables per inequality. *Math. Programming*, 62:69–83, 1993.

18. D. S. Hochbaum and D. B. Shmoys. A best possible approximation algorithm for the k-center problem. *Math. Oper. Res.*, 10:180–184, 1985.

19. D. S. Johnson. Approximation algorithms for combinatorial problems. *J. Comput. System Sci.*, 9:256–278, 1974.

20. A. A. Kuehn and M. J. Hamburger. A heuristic program for locating warehouses. *Management Sci.*, 9:643–666, 1963.

21. E. L. Lawler. *Combinatorial Optimization: Networks and Matroids*. Holt, Rinehart, and Winston, New York, 1976.

22. J. K. Lenstra, A. H. G. Rinnooy Kan, and P. Brucker. Complexity of machine scheduling problems. *Ann. Discrete Math.*, 1:343–362, 1977.

23. J.-H. Lin and J. S. Vitter. Approximation algorithms for geometric median problems. *Inform. Proc. Lett.*, 44:245–249, 1992.

24. J.-H. Lin and J. S. Vitter. ϵ-approximations with minimum packing constraint violation. In *Proceedings of the 24th Annual ACM Symposium on Theory of Computing*, pages 771–782, 1992.

25. L. Lovász. On the ratio of optimal integral and fractional covers. *Discrete Math.*, 13:383–390, 1975.

26. A. S. Manne. Plant location under economies-of-scale-decentralization and computation. *Management Sci.*, 11:213–235, 1964.

27. F. Margot, M. Queyranne, and Y. Wang. Decompositions, network flows and a precedence constrained single machine scheduling problem. Unpublished manuscript, December, 1996.

28. A. Munier and J. C. König. A heuristic for a scheduling problem with communication delays. *Oper. Res.*, 45:145–147, 1997.

29. C. A. Phillips, C. Stein, and J. Wein. Minimizing average completion time in the presence of release dates. *Math. Programming B*, 1998. To appear.

30. C. N. Potts. An algorithm for the single machine sequencing problem with precedence constraints. *Math. Programming Stud.*, 13:78–87, 1980.

31. M. Queyranne. Structure of a simple scheduling polyhedron. *Math. Programming*, 58:263–285, 1993.

32. P. Raghavan and C. D. Thompson. Randomized rounding: a technique for provably good algorithms and algorithmic proofs. *Combinatorica*, 7:365–374, 1987.

33. R. Ravi, A. Agrawal, and P. Klein. Ordering problems approximated: single-processor scheduling and interval graph completion. In *Proceedings of the 18th International Colloquium on Automata, Languages, and Processing, Lecture Notes in Computer Science 510*, pages 751–762, 1991.

34. A. S. Schulz and M. Skutella. Personal communication, 1997.

35. A. S. Schulz. *Scheduling and Polytopes*. PhD thesis, Technical University of Berlin, 1996.
36. D. B. Shmoys, É. Tardos, and K. I. Aardal. Approximation algorithms for facility location problems. In *Proceedings of the 29th Annual ACM Symposium on Theory of Computing*, pages 265–274, 1997.
37. J. B. Sidney. Decomposition algorithms for single-machine sequencing with precedence and deferral costs. *Oper. Res.*, pages 283–298, 1975.
38. J. F. Stollsteimer. *The effect of technical change and output expansion on the optimum number, size and location of pear marketing facilities in a California pear producing region*. PhD thesis, University of California at Berkeley, Berkeley, California, 1961.
39. J. F. Stollsteimer. A working model for plant numbers and locations. *J. Farm Econom.*, 45:631–645, 1963.
40. M. Sviridenko. Personal communication, July, 1997.
41. L. A. Wolsey. Mixed integer programming formulations for production planning and scheduling problems. Invited talk at the 12th International Symposium on Mathematical Programming, MIT, Cambridge, August, 1985.

The Steiner Tree Problem and Its Generalizations

Vijay V. Vazirani[1]

College of Computing, Georgia Institute of Technology, vazirani@cc.gatech.edu

Abstract. We will survey recent approximation algorithms for the metric Steiner tree problem and its generalization, the Steiner network problem. We will also discuss the bidirected cut relaxation for the metric Steiner tree problem.

1 Introduction

The Steiner tree problem occupies a central place in the emerging theory of approximation algorithms – methods devised to attack it have led to fundamental paradigms for the rest of the area. The reason for interest in this problem lies not only its rich mathematical structure, but also because it has arisen repeatedly in diverse application areas.

In the last couple of years, some nice algorithmic results have been obtained for this problem and its generalizations. Let us mention three that especially stand out: Arora's polynomial time approximation scheme [2] for the Euclidean Steiner tree problem, Promel and Steger's [18] factor $\frac{5}{3} + \epsilon$ approximation algorithm, for any constant $\epsilon > 0$, for the metric Steiner tree problem, and Jain's [12] factor 2 approximation algorithm for the Steiner network problem. Even though the Euclidean Steiner tree problem now seems fairly well understood (see also Du and Huang's [5] remarkable proof resolving the Gilbert-Pollack conjecture), it is clear that there are vast gaps in our understanding of the metric Steiner tree problem and its variants and generalizations.

In this survey, we will restrict attention to the metric case, and will first outline the ideas behind the algorithms of Promel and Steger, and Jain. Then, we will mention what is perhaps the most compelling open problem in this area: to design an algorithm using the bidirected cut relaxation for the metric Steiner tree problem, and determine the integrality gap of this relaxation.

2 Steiner trees via matroid parity

The *metric Steiner tree* problem is: Given a graph $G = (V, E)$ whose vertices are partitioned into two sets, R and S, the *required* and *Steiner* vertices, and a function cost : $E \to Q^+$ specifying non-negative costs for the edges, find a minimum cost tree containing all the required vertices and any subset of the Steiner vertices. It is easy to see that we can assume without loss of generality that the edge costs satisfy the triangle inequality.

Let us say that a Steiner tree is *3-restricted* if every Steiner vertex used in this tree has exactly three neighbors all of which are required vertices. Zelikovsky [21] showed that the cost of an optimal 3-restricted Steiner tree is within 5/3 of the cost of an optimal Steiner tree. Promel and Steger have shown how to find a 3-restricted Steiner tree that is within a $1 + \epsilon$ factor of an optimal such tree, for any $\epsilon > 0$. This gives a $5/3 + \epsilon$ factor approximation algorithm for the metric Steiner tree problem, for any $\epsilon > 0$.

A *hypergraph* $H = (V, F)$ is a generalization of a graph, allowing F to be an arbitrary family of subsets of V, instead of just 2 element subsets. A sequence of distinct vertices and hyperedges, $v_1, e_1, \ldots, v_l, e_l$, for $l \geq 2$ is said to be a *cycle* in H if $v_1 \in e_1 \cap e_l$ and for $2 \leq i \leq l$, $v_i \in e_{i-1} \cap e_i$. A subgraph of H, $H' = (V, F')$, with $F' \subseteq F$ is said to be a *spanning tree* of H if it is connected, acyclic and spans all vertices of V. Hypergraph H is said to be *3-regular* if every hyperedge in F consists of 3 vertices.

Consider an instance G of the metric Steiner tree problem. For a set of three required vertices and a single Steiner vertex, define their *connection cost* to be sum of the costs of the three edges connecting the Steiner vertex to each of the three required vertices. Now, define a hypergraph $H = (R, F)$ on the set of required vertices of G with edge costs as follows: F contains all edges of G incident at required vertices with their specified costs. In addition, for each triple of required vertices, F has a hyperedge on these vertices; the cost of this hyperedge is the minimum connection cost of these three vertices using some Steiner vertex.

Lemma 1. *A minimum cost spanning tree in H corresponds to an optimal 3-restricted Steiner tree in G.*

Lemma 2. *The problem of finding a minimum cost spanning tree in H can be reduced to that of finding a minimum cost spanning tree in a 3-regular hypergraph.*

The key step in Promel and Steger's result is:

Lemma 3. *The problem of finding a minimum cost spanning tree in a 3-regular hypergraph can be reduced to the minimum weight matroid parity problem.*

Let us sketch the reduction in Lemma 3. Let $H = (V, F)$ be a 3-regular hypergraph. A new graph H' on vertex set V is constructed as follows: corresponding to each hyperedge $\{v_1, v_2, v_3\} \in F$, we add the edge pair (v_1, v_2) and (v_1, v_3) to H' (the choice of v_1 is arbitrary). The cost of this pair is the same as that of the hyperedge. We will consider the graphic matriod in H'. It is easy to verify that a solution to the minimum weight matroid parity problem on this instance gives a minimum cost spanning tree in H.

Interestingly enough, determining the complexity of minimum weight matroid parity is still open, even though the cardinality version of this problem is known to be in P[15]. However, if the weights are given in unary, a random polynomial time algorithm is known [16] (see also [4]). Now, by scaling the original weights appropriately, we get a $1 + \epsilon$ factor algorithm for the minimum weight 3-restricted Steiner tree problem. The approximation algorithm for the metric Steiner tree problem follows.

For other algorithms for this problem see [3, 13]. An algorithm achieving a slightly better approximation factor of 1.644 appears in [14]. However, it is too involved in its current form for this survey; moreover, to beat the factor of 5/3, it takes time exceeding $O(n^{20})$.

3 Steiner networks via LP-rounding

The Steiner network problem generalizes the metric Steiner tree problem to higher connectivity requirements: Given a graph $G = (V, E)$, a cost function on edges $c : E \rightarrow \mathbf{Q}^+$ (not necessarily satisfying the triangle inequality), and a connectivity requirement function r mapping unordered pairs of vertices to $\mathbf{Z}^+$ find a minimum cost graph that has $r(u, v)$ edge disjoint paths for each pair of vetices $u, v \in V$. Multiple number of copies of any edge can be used to construct this graph; each copy of edge e will cost $c(e)$. For this purpose, for each edge $e \in E$, we are also specified $u_e \in \mathbf{Z}^+ \cup \{\infty\}$ stating an upper bound on the number of copies of edge e we are allowed to use; if $u_e = \infty$, then there is no bound on the number of copies of edge e.

All LP-duality based approximation algorithms for the metric Steiner tree problem and its generalizations work with the *undirected relaxation* [1, 9, 10, 20]. In order to give the integer programming formulation on which this relaxation is based, we will define a cut requirement function $f : 2^V \rightarrow \mathbf{Z}^+$. For $S \subseteq V$, $f(S)$ is defined to be the largest connectivity requirement separated by the cut $(S, \overline{S})$, i.e., $f(S) = \max\{r(u, v) | u \in S \text{ and } v \in \overline{S}\}$. Let us denote the set of edges in the cut $(S, \overline{S})$ by $\delta(S)$. The integer program has a variable x_e for each edge e:

$$\text{minimize} \quad \sum_{e \in E} c_e x_e \tag{1}$$

$$\text{subject to} \quad \sum_{e : \, e \in \delta(S)} x_e \geq f(S), \quad S \subseteq V$$

$$x_e \in \mathbf{Z}^+, \quad e \in E \text{ and } u_e = \infty$$

$$x_e \in \{0, 1, \ldots, u_e\}, \quad e \in E \text{ and } u_e \neq \infty$$

The LP-relaxation is:

$$\text{minimize} \quad \sum_{e \in E} c_e x_e \tag{2}$$

$$\text{subject to} \quad \sum_{e : \, e \in \delta(S)} x_e \geq f(S), \quad S \subseteq V$$

$$x_e \geq 0, \quad e \in E \text{ and } u_e = \infty$$

$$u_e \geq x_e \geq 0, \quad e \in E \text{ and } u_e \neq \infty$$

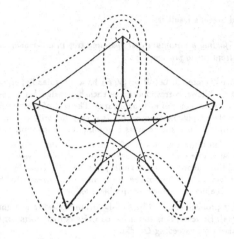

Figure 1 An extreme optimal solution for the Petersen graph.

Certain NP-hard problems, such a vertex cover [17] and node multiway cut [7] admit LP-relaxations having the remarkable property that they always have a half-integral optimal solution. Clearly, rounding up all halves to 1 in such a solution leads to a factor 2 approximation algorithm. Does the relaxation (2) have this property? The answer is "No". Not surprisingly, the Petersen graph is a counter-example: Consider the minimum spanning tree problem on this graph, i.e., for each pair of vertices u, v, $r(u, v) = 1$. Each edge is of unit cost. Since the Petersen graph is 3-edge connected (in fact, it is 3-vertex connected as well), $x_e = 1/3$ for each edge e is a feasible solution. Moreover, this solution is optimal, since the degree of each vertex under this solution is 1, the minimum needed to allow the connectivity required. The cost of this solution is 5. A half integral solution of cost 5 would have to pick, to the extent of half each, the edges of a Hamiltonian cycle. Since the Petersen graph has no Hamiltonian cycles, there is no half integral optimal solution.

Let us say that an *extreme solution*, also called a vertex solution or a basic feasible solution, for an LP is a feasible solution that cannot be written as the convex combination of two feasible solutions. It turns out that the solution, $x_e = 1/3$ for each edge e, is not an extreme solution. An extreme solution is shown in Figure 1; thick edges are picked to the extent of 1/2, thin edges to the extent of 1/4, and the missing edge is not picked. The isomorphism group of the Petersen graph is edge-transitive, and so there are several related extreme solutions; the solution $x_e = 1/3$ for each edge e is a suitable convex combination of these. Notice that although the extreme solution is not half-integral, it picks some edges to the extent of half.

Jain's algorithm is based on proving that in fact an extreme solution to LP (2) must pick at least one edge to the extent of at least a half. We will pay a factor of at most 2 in rounding up all such edges. But now how do we proceed? Let us start by computing the residual cut requirement function. Suppose H is the set of edges picked so far. Then, the residual requirement of cut $(S, \overline{S})$ is

$$f'(S) = \max\{f(S) - |\delta_H(S)|, 0\},$$

where $\delta_H(S)$ represents the set of edges of H crossing the cut $(S, \overline{S})$. In general, the *residual cut requirement function*, f', may not correspond to the cut requirement function for a certain set of connectivity requirements. We will need the following definitions to characterize it:

Definition 4. Function $f : 2^V \to \mathbf{Z}^+$ is said to be *submodular* if $f(V) = 0$, and for every two sets $A, B \subseteq V$, the following two conditions hold:

1. $f(A) + f(B) \geq f(A - B) + f(B - A)$.
2. $f(A) + f(B) \geq f(A \cap B) + f(A \cup B)$.

Lemma 5. *For any graph G on vertex set V, the function $|\delta_G(.)|$ is submodular.*

Definition 6. Function $f : 2^V \to \mathbf{Z}^+$ is said to be *weakly supermodular* if $f(V) = 0$, and for every two sets $A, B \subseteq V$, at least one the following conditions holds:

$$- f(A) + f(B) \leq f(A - B) + f(B - A)$$
$$- f(A) + f(B) \leq f(A \cap B) + f(A \cup B)$$

The following is an easy consequence of the definitions:

Lemma 7. *Let H be a subgraph of G. If $f : 2^{V(G)} \to Z^+$ is a weakly supermodular function, then so is the residual cut requirement function f'.*

It is easy to see that the original cut requirement function is weakly supermodular; by Lemma 7, so is the residual cut requirement function. Henceforth, we will assume that the function f used in LP (2) is a weakly supermodular function. We can now state the central polyhedral fact proved by Jain in its full generality. This will enable us to design an iterative algorithm for the Steiner network problem.

Theorem 8. *Any extreme solution to LP (2) picks some edge e to the extent of at least a half, i.e., $x_e \geq 1/2$.*

The algorithm that we started to design above can now be completed: in each iteration, round up all edges picked to the extent of at least a half in an extreme optimal solution, and update the residual cut requirement function. The algorithm halts when the original cut requirement function is completely satisfied, i.e., the residual cut requirement function is identically zero. Using a max-flow subroutine, one can obtain a separation oracle for LP (2) for any residual cut requirement function f, and so an extreme optimal solution can be computed in polynomial time.

Let us sketch how Theorem 8 is proven. From polyhedral combinatorics we know that a feasible solution to a set of linear inequalities in $\mathbb{R}^m$ is an extreme solution iff it satisfies m linearly independent inequalities with equality. W.l.o.g. we can assume that in any optimal solution to LP (2), for each edge e, $0 < x_e < 1$ (since edges with $x_e = 0$ can be dropped from the graph, and those with $x_e = 1$ can be permanently picked and the cut requirement function updated accordingly). So, the tight inequalities of an extreme optimal solution to LP (2) must correspond to cut requirements of sets.

Theorem 9. *Corresponding to any extreme solution to LP (2) there is a collection of m linearly independent tight sets that form a laminar family.*

The extreme optimal solution shown in Figure 1 uses 14 edges; we have shown a collection of 14 sets as required by Theorem 9.

Finally, a counting argument establishes Lemma 10, which leads to Theorem 8.

Lemma 10. *For any extreme solution to LP (2) there is a tight set S with exactly two edge in the cut $(S, \bar{S})$.*

The *integrality gap* of a relaxation is the supremum of the ratio of costs of optimal integral and optimal fractional solutions. Its importance lies in the fact that it limits the approximation factor that an algorithm using this relaxation can achieve. As a consequence of the factor 2 approximation algorithm for the Steiner network problem, we also get that the integrality gap of the undirected relaxation is 2.

Previously, algorithms achieving guarantees of $2k$ [20] and $2H_k$ [10], where k is the largest requirement, were obtained for this problem.

4 The bidirected cut relaxation

The metric Steiner tree problem is a special case of the Steiner network problem in which all requirements are 0 or 1. The further restriction, when all vertices are required, is the minimum spanning tree problem, which is polynomial time solvable. It turns out that the integrality gap of the undirected relaxation, LP 2, is essentially 2 even for restriction. To prove a lower bound of $2 - \frac{2}{n}$ on the integrality gap, consider a cycle on n vertices, with all edges of unit cost. The optimal integral solution to the minimum spanning tree problem on this graph is to pick $n - 1$ edges for a cost of $n - 1$, but an optimal fractional solution picks each edge to the extent of a half, for a total cost of $n/2$.

Thus, the undirected relaxation has an integrality gap of 2 not only for as general a problem as the Steiner network problem, but also for the minimum spanning tree problem, a problem in P! Two fundamental questions arise:

- Is there an exact relaxation, i.e., with integrality gap 1, for the minimum spanning tree problem?
- Is there a tighter relaxation for the metric Steiner tree problem?

The two questions appear to be intimately related: The answer to the first question is "Yes". This goes back to the seminal work of Edmonds [6], giving a primal-dual schema based polynomial time algorithm for the even more general problem of finding a minimum branching in a directed graph.

A similar idea gives a remarkable relaxation for the metric Steiner tree problem: the bidirected cut relaxation. This relaxation is conjectured to have integrality gap close to 1; the worst example known, due to Goemans [8], has integrality gap of 8/7. However, despite the fact that this relaxation has been known for decades, no algorithms have been designed using it, and the only upper bound known on its integrality gap is the trivial bound of factor 2 which follows from the undirected relaxation. Recently, [19] have given a primal-dual schema based factor 3/2 approximation algorithm using this relaxation for the special class of quasi-bipartite graphs; a graph is *quasi-bipartite* if it contains no edges between two Steiner vertices.

We present below the bidirected cut relaxation, and leave the open problem of designing an approximation algorithm beating factor 2 using it.

4.1 The bidirected cut relaxation

First replace each undirected edge (u, v) of G by two directed edges $(u \to v)$ and $(v \to u)$ each of cost $\text{cost}(u, v)$. Denote the graph so obtained by $\mathbf{G} = (V, \mathbf{E})$. Pick an arbitrary vertex $r \in R$ and designate it to be the *root*. W.r.t. the choice of a root, a set $C \subset V$ will be called a *valid set* if C contains at least one required vertex and $\overline{C}$ contains the root. The following integer program is trying to pick a minimum cost collection of edges from $\mathbf{E}$ in such a way that each valid set has at least one out-edge. It is easy to see that an optimal solution to this program will be a minimum cost Steiner tree directed into r.

$$\text{minimize} \quad \sum_{e \in \mathbf{E}} \text{cost}(e) x_e \tag{3}$$

$$\text{subject to} \quad \sum_{e:\, e \in \delta(C)} x_e \geq 1, \quad \forall \text{ valid set } C$$

$$x_e \in \{0, 1\}, \qquad \forall e \in \mathbf{E}$$

The LP-relaxation of this integer program is called the bidirected cut relaxation for the metric Steiner tree problem. (Notice that there is no need to explicitly add the costraints upper-bounding variables x_e.)

$$\text{minimize} \quad \sum_{e \in \mathbf{E}} \text{cost}(e) x_e \tag{4}$$

$$\text{subject to} \quad \sum_{e:\, e \in \delta(C)} x_e \geq 1, \quad \forall \text{ valid set } C$$

$$x_e \geq 0, \qquad \forall e \in \mathbf{E}$$

It is easy to verify that the choice of the root does not affect the cost of the optimal solution to the IP or the LP. As usual, the dual is seeking a maximum cut packing.

References

1. A. Agrawal, P. Klein, and R. Ravi. When trees collide: An approximation algorithm for the generalized Steiner problem on networks. *SIAM Journal on Computing*, 24(3):440-456, June 1995.
2. S. Arora. Nearly linear time approximation schemes for Euclidean TSP and other geometric problems. In *38th Annual Symposium on Foundations of Computer Science*, pages 554-563, Miami Beach, Florida, 20-22 October 1997. IEEE.
3. P. Berman and V. Ramaiyer. Improved approximations for the Steiner tree problem. *J. Algorithms*, 17, 381-408, 1994.
4. P. Camerini, G. Galbiati, and F. Maffioli. Random pseudo-polynomial algorithms for exact matroid problems. *J. Algorithms* 13, 258-273, 1992.
5. D. Du and F. Hwang. A proof of Gilbert-Pollack's conjecture on the Steiner ratio. *Algorithmica* 7, 121-135, 1992.
6. J. Edmonds. Optimum branchings. *J. Res. Nat. Bur. Standards*, B71:233-240, 1967.
7. N. Garg, V. V. Vazirani, and M. Yannakakis. Approximation algorithms for multiway cuts in node-weighted and directed graphs. *Proc. 21th International Colloquium on Automata, Languages and Programming*, 1994.
8. M. Goemans. Personal communication, 1996.

9. M.X. Goemans and D.P. Williamson. A general approximation technique for constrained forest problems. *SIAM Journal on Computing*, 24(2):296–317, April 1995.
10. M.X. Goemans, A. Goldberg, S. Plotkin, D. Shmoys, E. Tados, and D.P. Williamson. Approximation algorithms for network design problems. *SODA*, 223-232, 1994.
11. M. Grötschel, L. Lovász, and A. Schrijver. The ellipsoid method and its consequences in combinatorial optimization. *Combinatorica*, 1(2):169–197, 1981.
12. K. Jain. A factor 2 approximation algorithm for the generalized steiner network problem. manuscript, 1998.
13. L. Kou, G. Markowsky, and L. Berman. A fast algorithm for Steiner trees. *Acta Informatica* 15, 141-145, 1981.
14. M. Karpinski and A. Zelikovsky. New approximation algorithms for the Steiner tree problem. *Electr. Colloq. Comput. Compl.*, TR95-030, 1995.
15. L. Lovász. The matroid matching problem. *Algebraic Methods in Graph Theory*, Colloquia Mathematica Societatis Janos Bolyai, Szeged (Hungary), 1978.
16. H. Narayanan, H. Saran, and V.V. Vazirani. Randomized parallel algorithms for matroid union and intersection, with applications to arboresences and edge-disjoint spanning trees. *SIAM. J. Comput.*, vol. 23, No. 2, 387-397, 1994.
17. G. L. Nemhauser and L. E. Trotter, Jr.. Vertex packing: structural properties and algorithms. *Mathematical Programming*, 8:232-248, 1975.
18. H. J. Prömel and A. Steger. RNC-approximation algorithms for the Steiner problem. In R. Reischuk and M. Morvan, editors, *Proceedings of the Symposium on Theoretical Aspects of Computer Science*, volume 1200 of *Lecture Notes in Computer Science*, pages 559–570. Springer-Verlag, Berlin, Germany, 1997.
19. S. Rajagopalan and V.V. Vazirani. On the Bidirected Cut Relaxation for the Metric Steiner Tree Problem. Submitted, 1998.
20. D. Williamson, M. Goemans, M. Mihail, and V. Vazirani. A primal-dual approximation algorithm for generalized steiner network problems. *Combinatorica*, 15, 1995.
21. A. Zelikovsky. An 11/6-approximation algorithm for the network Steiner problem. *Algorithmica*, 9:463–470, 1993.

Approximation Schemes for Covering and Scheduling in Related Machines

Yossi Azar[1], Leah Epstein[2]

[1] Dept. of Computer Science, Tel-Aviv University. ***
[2] Dept. of Computer Science, Tel-Aviv University. †

Abstract. We consider the problem of assigning a set of jobs to m parallel related machines so as to maximize the minimum load over the machines. This situation corresponds to a case that a system which consists of the m machines is alive (i.e. productive) only when all machines are alive, and the system should be maintained alive as long as possible. The above problem is called related machines covering problem and is different from the related machines scheduling problem in which the goal is to minimize the maximum load. Our main result is a polynomial approximation scheme for this covering problem. To the best of our knowledge the previous best approximation algorithm has a performance ratio of 2. Also, an approximation scheme for the special case of identical machines was given by [14].
Some of our techniques are built on ideas of Hochbaum and Shmoys [12]. They provided an approximation scheme for the well known related machines scheduling. In fact, our algorithm can be adapted to provide a simpler approximation scheme for the related machines scheduling as well.

1 Introduction

We consider the problem of assigning a set of jobs to m parallel related machines so as to maximize the minimum load over the machines. This situation is motivated by the following scenario. A system which consists of m related machines is alive (i.e. productive) only when all machines are alive. The duration that a machine is alive is proportional to the total size of the resources (e.g. tanks of fuel) allocated to it. The goal is to keep the system alive as long as possible using a set of various sizes resources. The above problem has applications also in sequencing of maintenance actions for modular gas turbine aircraft engines [8]. To conform with the standard scheduling terminology we view the resources as jobs. Thus, jobs are assigned to machines so as to maximize the minimum load. In the related machines case each machine has its own speed

---

*** E-Mail: azar@math.tau.ac.il. Research supported in part by the Israel Science Foundation and by the United States-Israel Binational Science Foundation (BSF).
† E-Mail: lea@math.tau.ac.il.

(of the engine that operates on fuel). The identical machines case is a special case where all speeds of machines are identical. We refer to these problems as machines covering problems. Note that the classical scheduling/load-balancing problems [10, 11, 12, 13] seem strongly related to the covering problems. However, scheduling/load-balancing are packing problems and hence their goal is to minimize the makespan (i.e. to minimize the maximum load over all machines) where in the covering problems the goal is to maximize the minimum.

Our results: Our main result is polynomial approximation scheme for the covering problem in the related machines case. That is for any $\varepsilon > 0$ there is a polynomial time algorithm A_ε that approximates the optimal solution up to a factor of $1+\varepsilon$. In fact, since the problem is strongly NP hard no fully polynomial approximation scheme exists unless P=NP. Some of our techniques are built on ideas of Hochbaum and Shmoys [12]. They provided an approximation scheme for the well known related machines scheduling. In fact, our algorithm can be adapted to provide a simpler approximation scheme for the related machines scheduling as well.

Known results: The problem of maximizing the load of the least loaded machine, (i.e the machines covering problem) is known to be NP-complete in the strong sense already for identical machines [9]. For the identical machines case Deuermeyer, Friesen and Langston [7] studied the LPT-heuristic. The LPT-heuristic orders the jobs by non increasing weights and assigns each job to the least loaded machine at the current moment. It is shown in [7] that the approximation ratio of this heuristic is at most $\frac{4}{3}$. The tight ratio $\frac{4m-2}{3m-1}$ is given by Csirik, Kellerer and Woeginger [5]. Finally, Woeginger [14] designed a polynomial time approximation scheme for the identical machines covering problem.

The history for the *related* machines covering problem is much shorter. The only result which is known for the related machines case is the $2+\varepsilon$ approximation algorithm follows from [4]. The above paper also contains results for the on-line machines covering problems.

Definitions: We give a formal definition of the problems discussed above. Consider a set of m identical machines and a set of jobs. Machine i has a speed v_i and a job j has a weight w_j. The load of a machine i is the sum of the weights of the jobs assigned to it normalized by the speed. That is, $\ell_i = \sum_{j \in J_i} \frac{w_j}{v_i}$ where J_i is the of jobs assigned to machine i. (The identical machines case is the special case where all v_i are equal.) The goal in the machines covering problems is to assign the jobs to the machines so as to maximize the minimum load over the machines. This is in contrast to the scheduling/load-balancing problems where the goal there is to minimize the maximum load. Note that these covering problems are also different from the bin covering problems [1, 2, 3, 6] where the goal is to maximize the number of covered bins, i.e. bins of load of at least 1.

2 Approximation scheme for machine covering

We use a standard binary search technique to search for the value of the optimal cover. By this we reduce the approximation algorithm to the following approximate decision problem. Given a value T for the algorithm, the decision procedure outputs an assignment of value at least $(1 - \varepsilon)T$, or answers that there does not exist an assignment of value at least T. We start the binary search with an initial estimation of the value of the optimal cover using the $(2 + \varepsilon)$-approximation algorithm given in [4]. Clearly, the overall complexity of the approximation algorithm is just $O(\log 1/\varepsilon)$ times the complexity of the decision procedure (the initial estimation algorithm is fast).

We note that the decision procedure is equivalent to decide if one can fill the m bins such that bin i is filled by at least $(1 - \varepsilon)Tv_i$. We scale the sizes of bins and the weights of jobs, so that the size of the smallest bin is 1. Now we have a set of n jobs, and a set of m bins of sizes $s_1, s_2, ..., s_m$ where $1 = s_1 \leq s_2 \leq ... \leq s_m$.

Bin ranges: We partition the bins according to their sizes into sets B_r, where the bin range set B_r is the set of all bins of size $2^r \leq s_j < 2^{r+1}$. Let $R = \{r | B_r \neq \phi\}$, clearly $|R| \leq m$. We choose ε_0 to be a value such that $\frac{\varepsilon}{16} \leq \varepsilon_0 \leq \frac{\varepsilon}{8}$ and $\frac{1}{\varepsilon_0}$ is an integer. We denote $\varepsilon_r = 2^r \varepsilon_0$, $\delta_0 = \varepsilon_0^2$ and $\delta_r = 2^r \delta_0$. For each bin range B_r the jobs are partitioned into three sets.

- Big jobs: jobs of weight more than 2^{r+1}.
- Medium jobs: jobs of weight w_j: $\varepsilon_r < w_j \leq 2^{r+1}$.
- Small jobs: jobs of weight at most ε_r.

Jobs vectors: For each B_r we can approximate a set of jobs by a jobs vector $(y, n_1, n_2, \ldots, n_l, W)$ where y is the number of big jobs, n_k is the number of (medium) jobs whose size is between t_{k-1} and t_k where $t_k = \varepsilon_r + k\delta_r$ and W is the total weight of small jobs in the set. Clearly l, the number of types of medium jobs, is at most $\frac{2}{\delta_0} - \frac{1}{\varepsilon_0} \leq \frac{2}{\delta_0}$. We refer to the values of t_k as rounded weights. Note that the jobs vector for a given set of jobs depends on the bin range.

Cover vectors: Let B_r be the bin range of bin j. A cover vector for bin j has the same form as the jobs vector except the last coordinate which is an integer that corresponds to the weight of small jobs normalized by ε_r. A vector $(y, n_1, n_2, \ldots, n_l, q)$ is a cover vector for bin j if

$$2^{r+1}y + \sum_{k=1}^{l} n_k t_k + q\varepsilon_r \geq s_j - \varepsilon_r = s_j(1 - \varepsilon_0) \,,$$

i.e., the sum of the rounded weights of the jobs in the vector is at least $1 - \varepsilon_0$ fraction of s_j.

Let T_j' be the set of cover vectors for a bin j. Let T_j be the set of minimal cover vectors with respect to inclusion i.e. a cover vector u is in T_j if for any other cover vector for bin j, u' the vector $u - u'$ has at least one negative coordinate. Since the minimum weight of a job is $\varepsilon_0 2^r$ and the size of the bin is at most 2^{r+1} then any minimal cover vector consists of at most $\frac{2}{\varepsilon_0}$ jobs (sum of coordinates). Clearly, we may use only minimal vector covers since any cover can be transformed to a minimal one by omitting some jobs.

The layer graph: We use a layer graph where each node of the graph is state vector which is in a form of a cover vector. We order the bins in non decreasing size order. The layers of the graph are partitioned into phases. There are $|R|$ phases, a phase for each $r \in R$. Let $b_r = |B_r|$ for $r \in R$ and $b_r = 0$ otherwise. Phase r consists of $b_r + 1$ layers $L_{r,0}, ..., L_{r,b_r}$. The nodes of $L_{r,i}$ are all admissible state vectors of B_r. We put an edge between a node x' in $L_{r,i-1}$, and a node x in $L_{r,i}$ if the difference $u' = x' - x$ is a minimal cover vector of bin j.

Layer $L_{r,i}$ corresponds to jobs that remained after the first $j = \sum_{t=0}^{r-1} b_t + i$ bins were covered. More specifically, if there is a path from $L_{0,0}$ to a state vector in $L_{r,i}$ then there is a cover of the first j bins, where bin $k \leq j$ is covered with $s_k(1 - \varepsilon_0)$, such that the remaining jobs has jobs vector which is identical to the state vector except of the last coordinate. The last coordinate of the jobs vector, W, satisfies $q\varepsilon_r \leq (W + 2\varepsilon_r)(1 + \varepsilon_0)$ where q is the last coordinate of the state vector. Moreover, if there is a cover of the first j bins, bin $k \leq j$ with s_k such that the remaining jobs set defines some jobs vector then there a path from the first layer to the node in layer $L_{r,i}$ whose state vector is identical to the jobs vector except of the last coordinate. The last coordinate of the state vector, q, satisfies $W \leq q\varepsilon_r$ where W is the last coordinate of the jobs vector.

The translating edges between phases: The edges between phases "translate" each state vector into a state vector of the next phase. These edges are not used to cover bins, but to move from one phase to another. There is only one outgoing edge from each node in a last layer of any phase (except the last one which has no outgoing edges). More specifically, for each phase r, any node in L_{r,b_r} translates by an edge into a node in $L_{r',0}$ where $r' > r$ is the index of the next phase.

We consider a state vector in L_{r,b_r} (an **input vector**). We construct a corresponding state vector in layer $L_{r',0}$ (an **output vector**) which results in an edge between them. We start with an empty output state vector. We scan the input state vector. To build the output vector we need to know how jobs change their status. Since the bins become larger, the small jobs remain small. Medium jobs may either become small, or stay medium. Big jobs can stay big, or become medium or small.

First we consider the number of big items y in the input vector. Note that all big jobs could be used to cover previous bins. Thus we may assume that the smallest big jobs were used, and the y big jobs that remained are the largest y jobs in the original set. Let y_1 be the number of big jobs in the input vector that are also big in $B_{r'}$, y_2 the number of jobs that are medium in $B_{r'}$ and y_3 the number of jobs that are small in $B_{r'}$.

A medium job in phase B_r becomes small in $B_{r'}$ if its rounded weight t_j is at most $\varepsilon_{r'}$. In phase B_r the rounded weight of a job of weight $\varepsilon_{r'}$ is exactly $\varepsilon_{r'}$ since $(2^{r'-r} - 1)/\varepsilon_0$ is integer, and $\varepsilon_{r'} = 2^{r'}\varepsilon_0 = 2^r\varepsilon_0 + ((2^{r'-r} - 1)/\varepsilon_0 \cdot 2^r\varepsilon_0^2)$. Thus all medium jobs with $k \leq (2^{r'-r} - 1)/\varepsilon_0$ become small, and all other medium jobs remain medium.

The coordinates of the output vector: The big job coordinate in the output vector would be y_1, since no medium or small jobs could become big.

Now we deal with all the jobs which became small in the output, i.e. the y_3

big jobs together with the medium input jobs that became small output jobs and all the small jobs which must remain small in the output. To build the component of small jobs we re-estimate the total weight of small jobs. Since small jobs remain small, we initialize $W' = q\varepsilon_r$ where q is the integer value of the small jobs in the input node. We add to W' the total sum of the rounded jobs that were medium in B_r and become small in $B_{r'}$ (their rounded weight in B_r), and also the original weight of the big jobs in B_r that become small in $B_{r'}$. The new component q' of small jobs is $\lceil W'/\varepsilon_{r'} \rceil$.

Next we consider all jobs that are medium in $B_{r'}$. There were y_2 big jobs in B_r that become medium in $B_{r'}$. For every such job we round its weight according to $B_{r'}$ and add one to the coordinate of its corresponding type in $B_{r'}$. What remains to consider is the coordinates of jobs that are medium for both B_r and $B_{r'}$. We claim that all jobs that are rounded to one type k in B_r cannot be rounded to different types $B_{r'}$. Thus we add the type k coordinate of the input vector to the corresponding coordinate of the output vector. To prove the claim we note that a job of type k satisfies $2^r \varepsilon_0 + (k-1)2^r \varepsilon_0^2 < w_j \leq 2^r \varepsilon_0 + k 2^r \varepsilon_0^2$. Let $l = r' - r$, in terms of $\varepsilon_{r'}$ and $\delta_{r'}$ we get that

$$\varepsilon_{r'} + \delta_{r'} \left(\frac{1}{2^l}(\frac{1-2^l}{\varepsilon_0} + k - 1) \right) < w_j \leq \varepsilon_{r'} + \delta_{r'} \left(\frac{1}{2^l}(\frac{1-2^l}{\varepsilon_0} + k) \right).$$

Since $\frac{1-2^l}{\varepsilon_0} + k - 1$ and $\frac{1-2^l}{\varepsilon_0} + k$ are integers, then the interval $(\frac{1}{2^l}(\frac{1-2^l}{\varepsilon_0} + k - 1), \frac{1}{2^l}(\frac{1-2^l}{\varepsilon_0} + k))$ does not contain an integer and thus all these jobs are rounded to the type $\varepsilon_{r'} + \delta_{r'} \lceil \frac{1}{2^l}(\frac{1-2^l}{\varepsilon_0} + k) \rceil$.

After we built the graph, we look for a path between a node of the first layer in the first phase which corresponds to the whole set of jobs, and any node in the last layer of the last phase (which means that we managed to cover all bins with the set of jobs, maybe some jobs were not used but it is possible to add them to any bin). If such path exists, the procedure answers "yes", and otherwise "no".

3 Analysis of the algorithm

In this section we prove the correctness of our algorithm and compute its complexity.

Theorem 1. *If a feasible cover of value T exists, then the procedure outputs a path that corresponds to a cover of at least $(1 - \varepsilon)T$. Otherwise, the procedure answers "no".*

Proof. We need to show that any cover of value at least T can be transformed into a path in the graph, and that a path in the graph can be transformed into a cover of at least $(1 - \varepsilon)T$.

Assume a cover of value T. We first transform it to a cover of a type that our algorithm is searching for. We assume that the cover of any bin is minimal, otherwise we just remove jobs. Next we scan the bins by their order in the layered graph (small to large). If the j'th bin was covered by a single job, we change the cover to get another feasible cover in the following way. We consider the smallest job that is at least as large as the j'th bin, and is not used to cover a bin with a smaller index. We change the places of the two big jobs and continue the scanning. The modified cover after each change is still feasible since bin j is still covered and weight of the jobs on the larger bin may only increase. We use the final cover to build a path in the graph. We show how to add a single edge to a path in each step. Since there is only one outgoing edge from each node of the last layer in each phase, we only need to show how to add edges between consecutive layers inside the phases, each such edge corresponds to some bin. We also assume inductively the following invariant. The weight of the small jobs in any state vector is at least the weight of the small jobs that has not been yet used in the cover (the cover for bins we already considered).

Consider the j'th bin. If the bin is covered with a big job then we choose an edge that corresponds to one big job. Otherwise the bin is covered by medium jobs and small jobs. We compute the following state vector. Each medium job adds one to the appropriate coordinate of the vector. Note that rounded weight of a medium job (as interpreted in the state vector) is at least as large as its weight. We also need to provide one coordinate for the small jobs used in covering bin j. For that we divide the total weight of the small jobs in the cover of that bin by ε_r where r is the index of the phase and take the floor. The sum of the rounded weights of the jobs of the vector we built is at least $s_j - \varepsilon_r$, and thus it is a cover vector, i.e., an edge in the j'th layer of the graph. the invariant on the small jobs remains true since the weight of the small jobs of the cover vector is less than that in the cover.

Now we show that a full path in the graph is changed into a cover of value $(1 - \varepsilon)T$. We build the cover starting from the smallest bin (the first edge of the path). We show how to build a cover for one bin using a cover vector. In the case that the cover vector of the edge corresponds to one big job, we add the smallest big job available to the cover. We replace a medium job of rounded weight $\varepsilon_r + k\delta_r$ by a job of weight in the interval $(\varepsilon_r + (k-1)\delta_r, \varepsilon_r + k\delta_r]$. Such a job must exist according to the way that the graph was built. We replace the small jobs coordinate in the following way: Let $j_1, ..., j_z$ where $w_{j_1} \leq ... \leq w_{j_z}$ be the subset of the small jobs that were not used to cover any of the bins that were already covered. Let z_1 be the index $1 \leq z_1 \leq z$ such that

$$(q' - 3)\varepsilon_r < \sum_{1 \leq i \leq z_1} w_{j_i} \leq (q' - 2)\varepsilon_r$$

(where q' is the small jobs coordinate of the cover vector). We add the jobs $j_1, ..., j_{z_1}$ to the cover. We prove the following Lemma in order to show that such an index exists.

Lemma 2. *Consider the node in phase r on the path up to which we built the cover. Denote by W_1 the total rounded weight of small jobs that were not used to cover any of the bins that are already covered. (The rounded weight of a small job is its rounded weight when it was last medium, and its real weight if it never* *was medium). The small jobs coordinate q of the state vector of the node satisfies the equation*

$$q\varepsilon_r \leq W_1 + 2\varepsilon_r$$

Proof. First we show the correctness of the lemma only for the first layer of each phase. We show the lemma by induction on the number of phase r. In the beginning of each phase there is a rounding process in which the total rounded weight of small jobs is divided by ε_r, and the result is rounded up. Thus in phase 0 since we round up the number W'/ε_0, we added at most ε_0 to the total rounded weight of the jobs. In phase r we might add by rounding extra ε_r. Since $\sum_{r' < r} \varepsilon_{r'} < \varepsilon_r$ we conclude that the weights of the small jobs of a state vector in the first layer of a phase exceeds the total rounded weight of the small jobs available by at most $2\varepsilon_r$ and thus $q\varepsilon_r \leq W_1 + 2\varepsilon_r$. The lemma holds for nodes inside phases too, since each time we reduce q, by some number q_1, W_1 is reduced by at most $(q_1 - 2)\varepsilon$. This completes the proof of the lemma.

In a cover vector the small jobs coordinate q' is bounded above by the small jobs coordinate of the node for which this edge is outgoing. Thus $q'\varepsilon_r \leq q\varepsilon_r \leq W_1 + 2\varepsilon_r$ and thus $W_1 \geq (q' - 2)\varepsilon_r$. Since the weight of each job is at most ε_r, z_1 exists.

Now we show that this gives a cover of bin j. Let us calculate the ratio between the rounded weight and the real weight of jobs we assigned to this bin. Since we rounded only medium jobs, the loss in the weight of the job is at most δ_r, while its weight is at least ε_r. Thus the ratio is at most $(w_k + \delta_r)/w_k \leq 1 + \delta_r/\varepsilon_r \leq 1 + \varepsilon_0$. Thus the total weight of the jobs assigned to bin j is at least

$$(s_j - 4 \cdot \varepsilon_r)/(1 + \varepsilon_0) \geq (s_j - 4 \cdot 2^r \varepsilon_0)(1 - \varepsilon_0)$$
$$\geq s_j(1 - 4\varepsilon_0)(1 - \varepsilon_0) \geq s_j(1 - 5\varepsilon_0) \geq s_j(1 - \varepsilon)$$

where the second inequality follows from the fact that $s_j \geq 2^r$.

Complexity: The decision procedure consists of two parts.

1. Building the graph.
   The number of nodes in each layer of the graph is at most $N = 2(n + 2)n^{2/\varepsilon_0^2 + 1}$ since each of the coordinates of the first and the second parts of state vector does not exceed n, and the third one does not exceed $(n+2)(1 + \varepsilon_0) < 2(n + 2)$. Consider the a translation edges between layers of different phases. Recall that there is only one edge from each node. Now, consider the edges between layers that correspond to a bin. The number of edges from each node is at most $M = (2/\varepsilon_0^2 + 3)^{2/\varepsilon_0}$ (Since each of the $2/\varepsilon_0$ jobs can be one of $2/\varepsilon_0^2 + 2$ jobs types, or not to exist at all in the cover). Thus there are at most $mNM + mN$ edges in the graph. To build the edges inside phases we need to check at most M possibilities for each node inside a phase. It takes $O(\frac{1}{\varepsilon})$ to check one such possibility, i.e. that it corresponds to a minimal cover. Building one edge between phases takes $O(n)$ but there is only one outgoing edge from each node between phases. The total complexity of building the graph is $O(mNM/\varepsilon + mNn)$.

2. Running an algorithm to find a path, and translating it into a cover if it succeeds.

The algorithm of finding a path is linear in the number of edges. The complexity of building the cover is negligible.

The total complexity of the decision procedure is $O(c_\varepsilon n^{O(\frac{1}{\varepsilon})})$ where c_ε is a constant which depends only on ε. The complexity of the algorithm is clearly the complexity of the decision procedure times $O(\log 1/\varepsilon)$.

4 Approximation scheme for scheduling

It is easy to adapt the algorithm for machines covering and get a simpler algorithm for related machines scheduling. Here the big jobs cannot not be used for packing in bins and thus the algorithm becomes simpler. Of course, there are several straight forward changes such as rounding down the jobs instead of rounding them up and considering maximum packings instead of minimum covers. We omit the details.

References

1. N. Alon, J. Csirik, S. V. Sevastianov, A. P. A. Vestjens, and G. J. Woeginger. On-line and off-line approximation algorithms for vector covering problems. In *Proc. 4th European Symposium on Algorithms*, LNCS, pages 406–418. Springer, 1996.
2. S.F. Assmann. Problems in discrete applied mathematics. Technical report, Doctoral Dissertation, Mathematics Department, Massachusetts Institute of Technology, Cambridge, Massachusetts, 1983.
3. S.F. Assmann, D.S. Johnson, D.J. Kleitman, and J.Y.-T. Leung. On a dual version of the one-dimensional bin packing problem. *J. Algorithms*, 5:502–525, 1984.
4. Y. Azar and L. Epstein. On-line machine covering. In *5th Annual European Symposium on Algorithms*, pages 23–36, 1997.
5. J. Csirik, H. Kellerer, and G. Woeginger. The exact lpt-bound for maximizing the minimum completion time. *Operations Research Letters*, 11:281–287, 1992.
6. J. Csirik and V. Totik. On-line algorithms for a dual version of bin packing. *Discr. Appl. Math.*, 21:163–167, 1988.
7. B. Deuermeyer, D. Friesen, and M. Langston. Scheduling to maximize the minimum processor finish time in a multiprocessor system. *SIAM J. Discrete Methods*, 3:190–196, 1982.
8. D. Friesen and B. Deuermeyer. Analysis of greedy solutions for a replacement part sequencing problem. *Math. Oper. Res.*, 6:74–87, 1981.
9. M.R. Garey and D.S. Johnson. *Computers and Intractability*. W.H. Freeman and Company, San Francisco, 1979.
10. R.L. Graham. Bounds for certain multiprocessor anomalies. *Bell System Technical Journal*, 45:1563–1581, 1966.
11. R.L. Graham. Bounds on multiprocessing timing anomalies. *SIAM J. Appl. Math*, 17:263–269, 1969.
12. D. Hochbaum and D. Shmoys. A polynomial approximation scheme for scheduling on uniform processors: Using the dual approximation approach. *SIAM Journal on Computing*, 17(3):539–551, 1988.

13. D. S. Hochbaum and D. B. Shmoys. Using dual approximation algorithms for scheduling problems: Theoretical and practical results. *J. Assoc. Comput. Mach.*, 34(1):144–162, January 1987.
14. G. Woeginger. A polynomial time approximation scheme for maximizing the minimum machine completion time. *Operations Research Letters*, 20:149–154, 1997.

One for the Price of Two: A Unified Approach for Approximating Covering Problems *

Reuven Bar-Yehuda

Computer Science Dept, Tecnion IIT, Haifa 32000, Israel
`reuven@cs.technion.ac.il`,
WWW home page: `http://cs.technion.ac.il/~reuven`

Abstract. We present a simple and unified approach for developing and analyzing approximation algorithms for covering problems. We illustrate this on approximation algorithms for the following problems: Vertex Cover, Set Cover, Feedback Vertex Set, Generalized Steiner Forest and related problems.

The main idea can be phrased as follows: iteratively, pay two dollars (at most) to reduce the total optimum by one dollar (at least), so the rate of payment is no more than twice the rate of the optimum reduction. This implies a total payment (i.e., approximation cost) $\leq$ twice the optimum cost.

Our main contribution is based on a formal definition for covering problems, which includes all the above fundamental problems and others. We further extend the Bafna, Berman and Fujito Local-Ratio theorem. This extension eventually yields a short generic r-approximation algorithm which can generate most known approximation algorithms for most covering problems.

Another extension of the Local-Ratio theorem to randomized algorithms gives a simple proof of Pitt's randomized approximation for Vertex Cover. Using this approach, we develop a modified greedy algorithm, which for Vertex Cover, gives an expected performance ratio ≤ 2.

Keywords: Approximation Algorithm, Local Ratio, Covering Problems, Vertex Cover, Set Cover, Feedback Vertex Set, Generalized Steiner Forest, Randomized Approximations.

1 Introduction

Most algorithms for covering problems can be viewed according to two different approaches: the primal-dual principle, and the Local-Ratio principle (see, e.g., recent surveys [17],[19]). The common characteristic that we find in algorithms for covering problems is the use of vertex weight reductions (or vertex deletion,

---

* This research was supported by the fund for the promotion of research at the Technion.

which is a special case). We are interested in viewing algorithms from the viewpoint of the Local-Ratio theorem [4], and its extensions by Bafna, Berman and Fujito [2].

Following the work of Bafna et al., several studies have shown unified approaches for various families of covering problems. Fujito [13] gave a generic algorithm for node deletion. Chudack, Goemans, Hochbaum and Williamson [10] united algorithms for the Feedback Vertex Set problem presented by Bafna et al., [2] and by Becker and Geiger [9], using the primal-dual principle. Agrawal, Klein and Ravi [1] dealt with the Generalized Steiner Forest problem. Following their work, a couple of generalizations/unifications have recently been done, one by Williamson, Goemans, Mihail, and Vazirani [21] and another by Goemans and Williamson [15].

In this work, we try to combine all of these results into a unified definition for covering problems, and present a new local-ratio principle, which explains, with relative ease, the correctness of all the above algorithms. [1] Like the others, we use reductions of the weights in the graph, but we view this in the following way:

We try to measure the "effectiveness" of the reduction we are making; as a result of the reduction, there may be a change in the optimum. The reduction incurs a cost, and we would like the payment to be bounded by a constant r times the change in the problem's optimum.

This principle of "effectiveness" can also be viewed in terms of expected values, when the reduction we make is chosen according to some random distribution. This yields an extension of the local-ratio principle in another direction; in this extension, we are interested in bounding the expected change in the optimum as related to the expected cost.

This approach gives us a simple proof for Pitt's randomized approximation for Vertex Cover.

In addition, we get a new and interesting result: while a deterministic greedy algorithm [11] does not have a constant performance ratio, even for Vertex Cover, a randomized greedy algorithm has a performance ratio of 2.

1.1 Definitions

We are interested in defining a covering problem in such a way that it will contain the following problems: Vertex Cover, Set Cover, Feedback Vertex Set, Generalized Steiner Forest, Min 2-Sat, Min Cut, etc.

For a general covering problem, the input is a triple $(X, f : 2^X \to \{0, 1\}, \omega : X \to R^+)$, where X is a finite set, f is monotone, i.e., $A \subseteq B \Rightarrow f(A) \leq f(B)$, and $f(X) = 1$. For a set $C \subseteq X$, we define $\omega(C) = \sum_{x \in C} \omega(x)$, which we call the *weight* of C. We say that $C \subseteq X$ is a *cover* if $f(C) = 1$.

The objective of the problem is to find a cover $C^* \subseteq X$ satisfying

$$\omega(C^*) = \min\{\omega(C) | C \subseteq X \text{ and } f(C) = 1\}.$$

---

[1] For recent applications see [5] [6] [8].

Such a cover is called an *optimum cover*.

A cover $C \subseteq X$ is called an *r-approximation* if $\omega(C) \le r \cdot \omega(C^*)$, where C^* is an optimum cover. An algorithm A is called an *r-approximation* for a family of covering problems if, for every input (X, f, ω) in the family, A returns a cover which is an r-approximation.

1.2 Overview

This paper describes a unified approach for achieving r-approximations for covering problems. In Section 2 we describe the general idea of the technique, formalized by the Local-Ratio theorem. In Section 3 we show some applications of this technique to developing algorithms for the Vertex Cover problem. In Section 4 we apply the technique to Set Covering. In Section 5 we present an extension for dealing with randomized approximations. In Section 6 we describe an enhancement of the technique which allows us to deal with some recent approximations. In Section 7 we use the enhancement to describe and prove a 2-approximation algorithm for the Feedback Vertex Set problem, in Section 8 we use it for a 2-approximation for the Generalized Steiner Forest problem, and in Section 9 we present a generic r-approximation algorithm which can generate all deterministic algorithms presented in this paper.

2 First Technique: Pay for Every Weight Reduction

Given a triple (X, f, ω), our method for achieving an approximation cover is by manipulating the weight function ω. Denote the optimum cover of (X, f, ω) by $\mathrm{OPT}(\omega)$. The following is a fundamental observation for our technique:

The Decomposition Observation: For every two weight functions ω_1, ω_2,

$$\mathrm{OPT}(\omega_1) + \mathrm{OPT}(\omega_2) \le \mathrm{OPT}(\omega_1 + \omega_2)$$

We select some function $\delta : X \to R^+$, s.t. $\forall_{x \in X}\ 0 \le \delta(x) \le \omega(x)$

The function δ is used to reduce the weights, giving a new instance $(X, f, \omega - \delta)$. The value of $\mathrm{OPT}(\omega - \delta)$ may be smaller than $\mathrm{OPT}(\omega)$, and we denote their difference by $\Delta\mathrm{OPT} = \mathrm{OPT}(\omega) - \mathrm{OPT}(\omega - \delta)$. By the Decomposition Observation, $\Delta\mathrm{OPT} \ge \mathrm{OPT}(\delta)$.

This means that by reducing the weights by δ, we reduce the total optimum by at least $\mathrm{OPT}(\delta)$. How much are we willing to pay for this reduction in the optimum? Since our final objective is an r-approximation, we would like to pay no more than $r \cdot \mathrm{OPT}(\delta)$.

Let us be generous and pay for all weight reductions we make, i.e., define $\Delta\mathrm{COST}$ to be $\Delta\mathrm{COST} = \sum_{x \in X} \delta(x) = \delta(X)$. Therefore, what we need in order to get an r-approximation is

$$\delta(X) \le r \cdot \mathrm{OPT}(\delta)$$

or, in the terminology of the title: we are willing to pay no more than r times the gain in optimum reduction.

Let us now formalize the technique. Let (X, f, ω) be a covering problem triple. The function δ is called r-effective if:

1. $\forall_{x \in X} \ 0 \le \delta(x) \le \omega(x)$
2. $\delta(X) \le r \cdot \text{OPT}(\delta)$.

Now, consider the following general algorithm:

Algorithm $A(X, f, \omega)$

    Select r-effective δ
    Call algorithm $B(X, f, \omega - \delta)$ to find a cover C
    Return C

Theorem 1. The Local-Ratio Theorem:
If B returns a cover C s.t. $(\omega - \delta)(C) \le r \cdot OPT(\omega - \delta)$
then $\omega(C) \le r \cdot OPT(\omega)$

Proof. $\omega(C) =$
$$
\begin{aligned}
&= (\omega - \delta)(C) + \delta(C) && \text{[linearity of } \omega \text{ and } \delta] \\
&\le (\omega - \delta)(C) + \delta(X) && [\delta(C) \le \delta(X)] \\
&\le r \cdot OPT(\omega - \delta) + r \cdot OPT(\delta) && [B\text{'s property and the } r\text{-effectiveness of } \delta] \\
&\le r \cdot OPT(\omega) && \text{[Decomposition Observation]}
\end{aligned}
$$
$\square$

This is illustrated in the next section by a sequence of 2-approximation algorithms for the Vertex Cover problem.

3 The Vertex Cover Problem

Let $G = (V, E)$ be a simple graph with vertices V and edges $E \subseteq V^2$. The *degree* of a vertex $v \in V$ is defined by $d(v) = |\{e \in E : v \in e\}|$. The *maximum vertex degree* is defined by $\Delta_v = \max\{d(v) : v \in V\}$. A set $C \subseteq V$ is called a *vertex cover* of $G = (V, E)$ if every edge has at least one endpoint in C, i.e., $\forall_{e \in E} \ e \cap C \ne \phi$. The Vertex Cover problem (VC) is: given a simple graph $G = (V, E)$ and a weight $\omega(v) \ge 0$ for each $v \in V$, find a vertex cover C with minimum total weight. The Vertex Cover problem (VC) is NP-hard even for planar cubic graphs with unit weight [14]. For unit-weight Vertex Cover, Gavril (see [14]) suggested a linear time 2-approximation algorithm. For the general Vertex Cover problem, Nemauser and Trotter [18] developed a local optima algorithm that (trivially) implies a 2-approximation.

Until today, the best ratio known is 2. The first linear time algorithm was found by Bar-Yehuda and Even [3]. Their proof uses the primal-dual approach. It took a few more years to find a different kind of proof – the local-ratio theorem [4].

3.1 Bar-Yehuda & Even's 2-Approximation Algorithm

The main idea is as follows: let $\{x, y\} \in E$, and let $\epsilon = \min(\omega(x), \omega(y))$. Reducing $\omega(x)$ and $\omega(y)$ by ϵ has $\Delta\text{COST} = 2\epsilon$ while $\Delta\text{OPT} \geq \epsilon$, since one of the two vertices x, y participates in any optimum. So we pay 2ϵ in order to reduce the total optimum by at least ϵ, repeating this process until the optimum reaches zero. The total cost will be no more than twice the optimum.

Therefore, the 2-effective δ selected is

$$\delta(v) = \begin{cases} \epsilon & v \in \{x, y\} \\ 0 & \text{else} \end{cases}$$

Algorithm BarEven$(G = (V, E), \omega)$

For each $e \in E$
    Let $\epsilon = \min\{\omega(v) | v \in e\}$
    For each $v \in e$
        $\omega(v) = \omega(v) - \epsilon$
Return $C = \{v : \omega(v) = 0\}$

Theorem 2. *Algorithm BarEven is a 2-approximation.*

Proof. By induction on $|E|$.

The first step uses δ, which is 2-effective, since $\text{OPT}(\delta) = \epsilon$ and $\delta(V) = 2\epsilon$. The remaining steps perform a 2-approximation for $(G, \omega - \delta)$, by the induction hypothesis, and the Local Ratio Theorem. $\qquad\qquad\square$

For primal-dual proof see [3]. For first local-ratio proof, see [4].

3.2 Clarkson's Greedy Modification

Clarkson [12] uses the primal-dual approach to get the following 2-approximation algorithm.

Algorithm Clarkson $(G = (V, E), \omega)$

$C = \phi$
While $E \neq \phi$
    Let $x \in V$ with minimum $\epsilon = \frac{\omega(x)}{d(x)}$
    For each $y \in \Gamma(x)$ (the neighbors of x)
        $\omega(y) = \omega(y) - \epsilon$
    Add x to C and remove it from G
Return C

Theorem 3. *Clarkson's Algorithm is a 2-approximation.*

Proof. The δ that the algorithm chooses is $\delta(x) = \omega(x)$, $\delta(y) = \frac{\omega(x)}{d(x)}$ for every $y \in \Gamma(x)$, and 0 for all others. We need to show that δ is 2-effective, i.e. (1) $\forall_{v \in V}\ \delta(v) \leq \omega(v)$ (2) $\delta(V) \leq 2 \cdot \text{OPT}(\delta)$.

1. for $v = x$ we defined $\delta(x) = \omega(x)$. For $v \in \Gamma(x)$, we defined $\delta(v) = \frac{\omega(x)}{d(x)} \leq \frac{\omega(v)}{d(v)} \leq \omega(v)$. For all others, $\delta(v) = 0 \leq \omega(v)$.
2. $\delta(V) = \delta(x) + \delta(\Gamma(x)) + \delta(V - \{x\} - \Gamma(x)) = \omega(x) + \omega(x) + 0 = 2 \cdot \omega(x) = 2 \cdot \text{OPT}(\delta)$

The rest of the proof is by induction, using the Local Ratio Theorem. □

3.3 Subgraph Removal

Suppose we are given a family of graphs for which we wish to develop a 1.5-approximation. Getting rid of triangles is "free". To get rid of a triangle, say x, y, z, with weights $\epsilon = \omega(x) \leq \omega(y) \leq \omega(z)$, define $\delta(x) = \delta(y) = \delta(z) = \epsilon$, and $\forall_{v \notin \{x,y,z\}}\ \delta(v) = 0$. Obviously, $\text{OPT}(\delta) = 2\epsilon$, while $\delta(\{x, y, z\}) = 3\epsilon$, and therefore δ is 1.5-effective.

This was helpful, for example, in developing an efficient 1.5-approximation for planar graphs, see [4].

The $(2 - \log \log n/(2 \log n))$ - approximation algorithm [4] uses this approach to "get rid" of odd cycles of length $\leq 2k - 1$, where $k \approx 2 \log n/ \log \log n$. Let $C \subseteq V$ be a cycle of length $2k - 1$. Generalizing the idea for triangles, let $\epsilon = \min\{\omega(v) : v \in C\}$, and define $\delta(v) = \epsilon$ if $v \in C$, and 0 otherwise. In this case, $\delta(C)/\text{OPT}(\delta) \leq (2k - 1)/k = 2 - \frac{1}{k}$.

Bar-Yehuda and Even's linear time algorithm for VC can also be viewed as a subgraph removal algorithm, where the subgraph removed each time is an edge.

4 Vertex Cover in Hypergraphs

Let $G = (V, E)$ be a simple[2] hypergraph with vertices V and edges $E \subseteq 2^V$. The *degree of a vertex* $v \in V$ is defined by $d(v) = |\{e \in E : v \in e\}|$. The *maximum vertex degree* is defined by $\Delta_v = \max\{d(v) : v \in V\}$. The *degree of an edge* $e \in E$ is defined by $d(e) = |e|$. The *maximum edge degree* is defined by $\Delta_e = \max\{|e| : e \in E\}$. A set $C \subseteq V$ is called a *vertex cover* of $G = (V, E)$ if every edge $e \in E$ has at least one endpoint in C (i.e., $e \cap C \neq \phi$). The Hypergraph Vertex Cover problem (HVC) is: given $G = (V, E)$, and for each $v \in V$ a weight $\omega(v) \geq 0$, find a cover with minimum total weight.

This generalization of the VC problem is exactly the Set Covering problem (SC).

Algorithm BarEven is written in hypergraph terminology, and therefore approximates HVC as well.

Theorem 4. *Algorithm BarEven is a Δ_e-approximation for HVC.*

---

[2] A hypergraph is called simple if no edge is a subset of another edge.

Proof. For $e \in E, \epsilon = \min\{\omega(v) : v \in e\}$. Define $\delta(v) = \epsilon$ for all $v \in e$ and 0 otherwise.

Since $\mathrm{OPT}(\delta) = \epsilon$ and $\delta(V) = \delta(e) = |e| \cdot \epsilon \leq \Delta_e \cdot \epsilon$, this implies that δ is Δ_e-effective. $\square$

5 Randomized r-approximations

Let us select $\delta : X \to R^+$ from some random distribution. Such a random distribution is called r-effective if

1. $\forall_{x \in X}\ 0 \leq \delta(X) \leq \omega(x)$ and
2. $Exp[\delta(X)] \leq r \cdot Exp[\mathrm{OPT}(\delta)]$

Intuitively: the randomized rate of payment is no more than r times the randomized rate of the optimum reduction.

Now consider the following general algorithm:

Algorithm $A(X, f, \omega)$

Select $\delta : X \to R^+$ from some r-effective random distribution
Call algorithm $B(X, f, \omega - \delta)$ to find a cover C
Return C

Theorem 5. The Randomized Local-Ratio Theorem:
If B returns a cover C s.t. $Exp[(\omega - \delta)(C)] \leq r \cdot Exp[\mathrm{OPT}(\omega - \delta)]$, then $Exp[\omega(C)] \leq r \cdot OPT(\omega)$.

Proof. $Exp[\omega(C)] =$
$= Exp[(\omega - \delta)(C)] + Exp[\delta(C)]$
$\leq r \cdot Exp[\mathrm{OPT}(\omega - \delta)] + r \cdot Exp[\mathrm{OPT}(\delta)]$ [from the properties of B and
of the distribution]
$= r \cdot Exp[\mathrm{OPT}(\omega - \delta) + \mathrm{OPT}(\delta)]$
$\leq r \cdot Exp[\mathrm{OPT}(\omega)]$ [by the Decomposition Observation.]
$= r \cdot \mathrm{OPT}(\omega)$.

$\square$

An algorithm A that produces a cover C satisfying $Exp[\omega(C)] \leq r \cdot \mathrm{OPT}(\omega)$ is called a *randomized-r-approximation*.

5.1 Pitt's Randomized-2-Approximation Algorithm

Algorithm Pitt $(G = (V, E), \omega)$

$C = \{x : \omega(x) = 0\}$ and delete C from G
While $E \neq \phi$
    Select $e \in E$
    Let $\bar{\omega} = \dfrac{1}{\sum_{x \in e} \frac{1}{\omega(x)}}$
    Select at random $x \in e$ with $Prob(x) = \frac{\bar{\omega}}{\omega(x)}$
    Remove x from G and add it to C
Return C

Pitt [20] uses the primal-dual approach to prove that his algorithm is a randomized-2-approximation algorithm for Vertex Cover. Using our technique, we get:

Theorem 6. *Pitt's algorithm is a randomized-Δ_e-approximation for HVC.*

Proof. To prove correctness, we use the Randomized Local-Ratio theorem. We can regard the selection of a vertex x and its removal as if we defined $\delta(x) = \omega(x)$, and $\forall_{y \neq x} \ \delta(y) = 0$. We show that in each iteration, the distribution is Δ_e-effective. For this, we need to show that $Exp[\delta(V)] \leq r \cdot Exp[OPT(\delta)]$.

Let e be the edge chosen in the first iteration, and let C^* be an optimum cover.

$$Exp[OPT(\delta)] = Exp[\delta(C^*)] = \sum_{x \in e \cap C^*} p(x) \cdot \omega(x)$$

$$= \sum_{x \in e \cap C^*} \frac{\bar{\omega}}{\omega(x)} \cdot \omega(x) = \bar{\omega} \cdot |e \cap C^*| \geq \bar{\omega}$$

$$Exp[\delta(V)] = \sum_{x \in V} p(x) \cdot \omega(x) = \sum_{x \in e} \frac{\bar{\omega}}{\omega(x)} \cdot \omega(x) = \bar{\omega} \cdot |e| \leq \bar{\omega} \cdot \Delta_e$$

$\square$

5.2 Our Modified Greedy Algorithm

The greedy algorithm for set cover [11] is a $ln\Delta_v$-approximation. Algorithm BarEven is a Δ_e-approximation. In some families of problems (e.g., VC), Δ_e is preferable to $ln\Delta_v$. In these cases, we would like to have a ratio of Δ_e, but we may also prefer to use the "greedy" heuristic. Can we have the best of both worlds? As we have seen, this is possible: Clarkson's modified greedy algorithms for VC can easily be modified for HVC, using the Local-Ratio theorem. But, even for unit-weight Vertex Cover, it may have a problem with robustness, due to error accumulation. The following randomized greedy modification does not "reduce weights", but rather deletes vertices:

Algorithm Random-Greedy $(G = (V, E), \omega)$

$C = \{x : \omega(x) = 0\}$ and delete C from G
While $E \neq \phi$
    Let $\bar{\omega} = \dfrac{1}{\sum_{x \in V} \frac{d(x)}{\omega(x)}}$
    Select at random $x \in V$ with $Prob(x) = \frac{d(x)}{\omega(x)} \cdot \bar{\omega}$
    Remove x from G and add it to C
Return C

Let C^* be an optimum cover,

$$Exp[\text{OPT}(\delta)] = Exp[\delta(C^*)] = \sum_{x \in C^*} p(x) \cdot \omega(x) = \sum_{x \in C^*} d(x) \cdot \bar{\omega} \geq |E| \cdot \bar{\omega}$$

$$Exp[\delta(V)] = \sum_{x \in V} p(x) \cdot \omega(x) = \sum_{x \in V} d(x) \cdot \bar{\omega} \leq \Delta_e \cdot |E| \cdot \bar{\omega}$$

this implies:

Theorem 7. *Random-Greedy is a randomized-Δ_e-approximation for the HVC problem.*

Corollary 1. *Random-Greedy is a randomized-2-approximation for VC.*

6 Pay Only for Minimal Covers

Following Bafna, Berman and Fujito [2], we extend the Local-Ratio theorem. This will be shown to be useful in the following sections.

In the previous analyses, we used $\delta(V)$ as an upper bound for the payment in a step. This may be too much. If we knew the final cover C eventually found by the algorithm, we would know that $\delta(C)$ is the exact payment.

In practice, we can restrict the final cover C to be a minimal cover. Defined formally:

Definition A cover C is a *minimal cover* if $\forall_{x \in c} C \backslash \{x\}$ is not a cover. A cover C is a *minimal cover w.r.t a weight function* ω, if $\forall_{x \in C \text{ and } \omega(x) > 0} C \backslash \{x\}$ is not a cover.

If we know that the final cover C is minimal w.r.t. δ, we can bound the payment ΔCOST by $\max\{\delta(C) | C$ is minimal w.r.t $\delta\}$.

Define $\delta : X \to R^+$ to be *minimal-r-effective* if

1. $\forall_{x \in X} \ 0 \leq \delta(X) \leq \omega(x)$
2. $\delta(C) \leq r \cdot \text{OPT}(\omega)$ for all C that are minimal w.r.t. δ.

Consider the following general algorithm A:

Algorithm $A(X, f, \omega)$

Select a minimal-r-effective δ
Call algorithm $B(X, f, \omega - \delta)$ to find a cover C
Removal loop:
for each $x \in C$, if $\delta(x) > 0$ and $C \setminus \{x\}$ is a cover then $C = C \setminus \{x\}$
Return C

Theorem 8. The Extended Local-Ratio Theorem
If B returns a cover C s.t. $(\omega - \delta)(C) \leq r \cdot OPT(\omega - \delta)$, then $\omega(C) \leq r \cdot OPT(\omega)$.

Proof. After the "removal loop", C is still a cover, but it is also minimal w.r.t. δ, and therefore $\delta(C) \leq r \cdot OPT(\delta)$. Hence

$$
\begin{aligned}
\omega(C) &= (\omega - \delta)(C) + \delta(C) \\
&\leq r \cdot OPT(\omega - \delta) + r \cdot OPT(\delta) \quad \text{[by the properties of } B \text{ and } \delta] \\
&\leq r \cdot OPT(\omega) \quad\quad\quad\quad\quad\quad\quad \text{[by the Decomposition Observation]}
\end{aligned}
$$

$\square$

7 The Feedback Vertex Set Problem

Let $G = (V, E)$ be a simple graph with weight function $\omega : V \to R^+$. A set $F \subseteq V$ is called a Feedback Vertex Set (FVS) if $G \setminus F$ is a forest, i.e., for every cycle $C \subseteq V$, $C \cap F \neq \phi$. The FVS problem is: given a graph $G = (V, E)$ and a weight function $\omega : V \to R^+$, find a FVS, F, with minimum total weight. The FVS problem is NP-hard [14]. Furthermore, any r-approximation for FVS implies an r-approximation with the same time complexity for the VC problem (each edge in the VC instance graph can be replaced by some cycle).

This implies that a 2-approximation is the best we can hope for, as long as the ratio for VC is not improved.

The first non-trivial approximation algorithm was given by Bar-Yehuda, Geiger, Naor and Roth [7]. They present a 4-approximation for unit weights and an $O(\log n)$-approximation for the general problem. Following their paper, Bafna, Berman and Fujito [2] found the first 2-approximation. They were also the first to generalize the local ratio theorem to deal with "paying" according to a minimal cover. Their approach has helped to deepen our understanding of approximations for covering problems.

Following their approach, Becker and Geiger [9] present a relatively simple 2-approximation. Following all these, Chudak, Goemans, Hochbaum and Williamson [10] explained all of these algorithms in terms of the primal-dual method. They also gave a new 2-approximation algorithm, which is a simplification of the Bafna et al. algorithm. Using the Extended Local Ratio theorem, we can further simplify the presentation and proof of all these three algorithms, [2], [10], [9]. Let us illustrate this for the Becker-Geiger algorithm.

The main idea behind Becker and Geiger's algorithm is the weighted graph we call a degree-weighted graph. A *degree-weighted graph* is a pair (G, ω) s.t. each vertex v has a weight $\omega(v) = d(v) > 1$.

Theorem 9. *Every minimal FVS in a degree-weighted graph is a 2-approximation.*

Proof. Let C be a minimal cover, and C^* an optimum cover. We need to prove $\omega(C) \leq 2 \cdot \omega(C^*)$. It is enough to show that

$$\sum_{x \in C} \omega(x) \leq 2 \cdot \sum_{x \in C^*} \omega(x)$$

by definition, it is enough to show that

$$\sum_{x \in C} d(x) \leq 2 \cdot \sum_{x \in C^*} d(x).$$

This is proved in [9]. A simple proof also appears in [10]. $\square$

Corollary 2. *If $G = (V, E)$ is a graph with $\forall_{v \in V} \ d(v) > 1$, and $\delta : V \to R^+$ satisfies $\forall_{v \in V} \ 0 \leq \delta(v) = \epsilon \cdot d(v) \leq \omega(v)$ for some ϵ, then δ is 2-effective.*

Proof. We need only show that for every minimal cover C, $\delta(C) \leq 2 \cdot \text{OPT}(\delta)$. This is immediate from Theorem 9. $\square$

In order to guarantee termination, we choose $\epsilon = \min_{v \in V} \frac{\omega(v)}{d(v)}$.

Algorithm BeckerGeiger $(G = (V, E), \omega)$

If G is a tree return ϕ
If $\exists_{x \in V} \ d(x) \leq 1$ return BeckerGeiger$(G \backslash \{x\}, \omega)$
If $\exists_{x \in V} \ w(x) = 0$ return $\{x\}$+BeckerGeiger$(G \backslash \{x\}, \omega)$
Let $\epsilon = \min_{v \in V} \frac{\omega(v)}{d(v)}$
Define $\delta : \forall_{x \in V} \ \delta(x) = \epsilon \cdot d(x)$
$C = $ BeckerGeiger$(G, \omega - \delta)$
Removal loop:
for each $x \in C$, if $\delta(x) > 0$ and $C \backslash \{x\}$ is a cover then $C = C \backslash \{x\}$

Theorem 10. *Algorithm BeckerGeiger is a 2-approximation for FVC.*

Proof. The algorithm terminates, since in each recursive call, at least one more vertex weight becomes zero. Therefore we can use induction. The base of induction is trivial. Let us assume inductively that the recursive call returns C s.t.
$$(\omega - \delta)(C) \leq 2 \cdot \text{OPT}(\omega - \delta).$$

Since, by Corollary 2, δ is 2-effective, we can use the Extended Local-Ratio theorem to conclude that $\omega(C) \leq 2 \cdot \text{OPT}(\omega)$. $\square$

8 Generalized Steiner Forest and Related Problems

Goemans and Williamson [15] have recently presented a general approximation technique for covering cuts of vertices with edges. Their framework includes the Shortest Path, Minimum-cost Spanning Tree, Minimum-weight Perfect Matching, Traveling Salesman, and Generalized Steiner Forest problems. All these problems are called Constraint Forest Problems.

To simplify the presentation, let us choose the Generalized Steiner Forest problem. We are given a graph $G = (V, E)$, a weight function $\omega : E \to R^+$, and a collection of terminal sets $T = \{T_1, T_2, \ldots, T_t\}$, each a subset of V. A subset of edges $C \subseteq E$ is called a Steiner-cover if, for each T_i, and for every pair of terminals $x, y \in T_i$, there exists a path in C connecting x to y. The Generalized Steiner Forest problem is: given a triple (G, T, ω), find a Steiner-cover C with minimum total weight $\omega(C)$.

In order to get a 2-effective weight function, let us define an edge-degree-weighted graph.

An *edge-degree-weighted graph* is a triple (G, T, ω) s.t.

$$\forall_{e \in E} \ \omega(e) = d_T(e) = |\{x \in e : \exists_i \ x \in T_i\}| \ .$$

Theorem 11. *Every minimal cover in an edge-degree-weighted graph is a 2-approximation.*

Proof. The proof is an elementary exercise, see, e.g., [16]. □

So now, given (G, T, ω), we can easily compute a 2-effective weight function $\delta(e) = d_T(e) \cdot \epsilon$ for all $e \in E$, and in order to guarantee termination, we select $\epsilon = \min\{\omega(e)/d_T(e) | d_T(e) \neq 0\}$.

Now we can proceed as follows:

"Shrink" all pairs $e = \{x, y\}$ s.t. $(\omega - \delta)(e) = 0$.

Recursively, find a minimal 2-approximation Steiner-cover, C. Add all "shrunken" edges to the cover C. In order to guarantee the minimality property, apply the following removal loop:

For each "shrunken" edge e do: if $C \setminus \{x\}$ is a Steiner-cover, delete e from C

9 A Generic r-Approximation Algorithm

We now present a generic r-approximation algorithm that can be used to generate all the deterministic algorithms presented in this paper.

The weight function δ is called ω-tight if $\exists_{x \in X} \ \delta(x) = \omega(x) > 0$.

Algorithm Cover (X, f, ω)

If the set $\{x \in X : \omega(x) = 0\}$ is a cover then return this set.
Select ω-tight minimal-r-effective $\delta : X \to R^+$
Recursively call Cover $(X, f, \omega - \delta)$ to get a cover C
Removal loop:
for each $x \in C$, if $\delta(x) > 0$ and $C \backslash \{x\}$ is a cover then $C = C \backslash \{x\}$

Theorem 12. *Algorithm Cover is an r-approximation.*

Proof. Define $X^+ = \{x \in X | \omega(x) > 0\}$. Since δ is ω-tight, the size of X^+ decreases at each recursive call. This implies that the total number of recursive calls is bounded by $|X|$. We can now prove the theorem using induction on $|X^+|$.

Basis: $|X^+| = 0$, hence $\omega(X) = 0$.

Step: We can consider the recursive call as the algorithm B in the Extended Local-Ratio theorem, theorem 8. By the induction hypothesis, this recursive call, Cover $(X, f, \omega - \delta)$, satisfies B's requirement in the theorem, i.e., $(\omega - \delta)(X) \leq r \cdot \mathrm{OPT}(\omega - \delta)$. Since δ is minimal-r-effective, it remains to show that C is a minimal cover w.r.t δ. This follows from the last step of algorithm Cover. $\square$

Acknowledgment
We would like to thank Yishay Rabinovitch and Joseph Naor for helpful discussions, Avigail Orni for her careful reading and suggestions, and Yvonne Sagi for typing.

References

1. A. Agrawal, P. Klein, and R. Ravi. When trees collide: an approximation algorithm for the generalized steiner problem in networks. *Proc. 23rd ACM Symp. on Theory of Computing*, pages 134–144, 1991.
2. V. Bafna, P. Berman, and T. Fujito. Constant ratio approximation of the weighted feedback vertex set problem for undirected graphs. *ISAAC '95 Algorithms and Computation*, (1004):142–151, 1995.
3. R. Bar-Yehuda and S. Even. A linear time approximation algorithm for the weighted vertex cover problem. *Journal of Algorithms*, 2:198–203, 1981.
4. R. Bar-Yehuda and S. Even. A local-ratio theorem for approximating the weighted vertex cover problem. *Annals of Discrete Mathematics*, 25:27–46, 1985.
5. R. Bar-Yehuda. A linear time 2-approximation algorithm for the min clique-complement problem. *Technical Report CS0933, Technion Haifa*, May 1998.
6. R. Bar-Yehuda. Partial vertex cover problem and its generalizations. *Technical Report CS0934, Technion Haifa*, May 1998.
7. R. Bar-Yehuda, D. Geiger, J. Naor, and R. Roth. Approximation algorithms for the vertex feedback set problem with applications to constraint satisfaction and bayesian inference. *Accepted to SIAM J. on Computing*, 1997.
8. R. Bar-Yehuda and D. Rawitz. Generalized algorithms for bounded integer programs with two variables per constraint. *Technical Report CS0935, Technion Haifa*, May 1998.

9. A. Becker and D. Geiger. Approximation algorithms for the loop cutset problem. *Proc. 10th Conf. on Uncertainty in Artificial Intelligence*, pages 60–68, 1994.

10. F. Chudak, M. Goemans, D. Hochbaum, and D. Williamson. A primal-dual interpretation of recent 2-approximation algorithms for the feedback vertex set problem in undirected graphs. *Unpublished*, 1996.

11. V. Chvatal. A greedy heuristic for the set-covering problem. *Math. of Oper. Res.*, 4(3):233–235, 1979.

12. K. Clarkson. A modification of the greedy algorithm for the vertex cover. *Info. Proc. Lett.*, 16:23–25, 1983.

13. T. Fujito. A unified local ratio approximation of node-deletion problems. *ESA, Barcelona, Spain*, pages 167–178, September 1996.

14. M. Garey and D. Johnson. *Computers and Intractability*. W.H. Freeman, 1979.

15. M. Goemans and D. Williamson. A general approximation technique for constrained forest problems. *SIAM Journal on Computing*, 24(2):296–317, 1995.

16. M. Goemans and D. Williamson. The primal-dual method for approximation algorithms and its application to network design problems. *Approximation Algorithms for NP-Hard Problems*, 4, 1996.

17. D. Hochbaum. Approximating covering and packing problems: Set cover, vertex cover, independent set, and related problems. *Chapter 3 in Approximation Algorithms for NP-Hard Problems, PWS Publication Company*, 1997.

18. G. Nemhauser and J. L.E. Trotter. Vertex packings: structural properties and algorithms. *Mathematical Programming*, 8:232–248, 1975.

19. V. T. Paschos. A survey of approximately optimal solutions to some covering and packing problems. *ACM Computing Surveys*, 29(2):171–209, June 1997.

20. L. Pitt. Simple probabilistic approximation algorithm for the vertex cover problem. *Technical Report, Yale*, June 1984.

21. D. Williamson, M. Goemans, M. Mihail, and V. Vazirani. A primal-dual approximation algorithm for generalized steiner network problems. *Combinatorica*, 15:435–454, 1995.

Approximation of Geometric Dispersion Problems [*]

(Extended Abstract)

Christoph Baur and Sándor P. Fekete

Center for Parallel Computing, Universität zu Köln, D–50923 Köln, Germany
[baur,fekete]@zpr.uni-koeln.de

Abstract. We consider problems of distributing a number of points within a connected polygonal domain P, such that the points are "far away" from each other. Problems of this type have been considered before for the case where the possible locations form a discrete set. Dispersion problems are closely related to packing problems. While Hochbaum and Maass (1985) have given a polynomial time approximation scheme for packing, we show that geometric dispersion problems cannot be approximated arbitrarily well in polynomial time, unless P=NP. We give a $\frac{2}{3}$ approximation algorithm for one version of the geometric dispersion problem. This algorithm is strongly polynomial in the size of the input, i.e., its running time does not depend on the area of P. We also discuss extensions and open problems.

1   Introduction: Geometric Packing Problems

In the following, we give an overview over geometric packing. Problems of this type are closely related to geometric dispersion problems, which are described in Section 2.

Two-dimensional packing problems arise in many industrial applications. As two–dimensional cutting stock problems, they occur whenever steel, glass, wood, or textile materials are cut. There are also many other problems that can be modeled as packing problems, like the optimal layout of chips in VLSI, machine scheduling, or optimizing the layout of advertisements in newspapers.

When considering the problem of finding the best way to pack a set of objects into a given domain, there are several objectives that can be pursued: We can try to maximize the value of a subset of the objects that can be packed and consider *knapsack problems*; we can try to minimize the number of containers that are used and deal with *bin packing problems*, or try to minimize the area that is used – in *strip packing problems*, this is done for the scenario where the domain is a strip with fixed width and variable length that is to be kept small.

All of these problems are NP-hard in the strong sense, since they contain the one-dimensional bin packing problem as a special case. However, there are

---

[*] This work was supported by the German Federal Ministry of Education, Science, Research and Technology (BMBF, Förderkennzeichen 01 IR 411 C7).

some important differences between the one- and the two-dimensional instances; and while there are many algorithms for one-dimensional packing problems that work well in practice (currently, benchmark instances of the one-dimensional knapsack problem with up to 250,000 objects can be solved optimally, see [31]), until recently, the largest solved benchmark instance of the two-dimensional orthogonal knapsack problem (i.e., packing rectangles into a rectangular container) had no more than 23 objects (see [3, 22]). One of the difficulties in two dimensions arises from the fact that an appropriate way of modeling packings is not easy to find; this is highlighted by the fact that the feasible space cannot be assumed to be convex. (Even if the original domain is convex, the remaining feasible space will usually lose this property after a single object is placed in the domain.) This makes it impossible to use standard methods of combinatorial optimization without additional insights. For an overview over heuristic and exact packing methods, see [40]. See [11–14, 38] for a recent approach that uses a combination of geometric and graph-theoretic properties for characterizing packings of rectangles and constructing relatively fast exact algorithms. Kenyon and Remila [24] give an "asymptotic" polynomial time approximation scheme for the strip packing problem, using the additional assumption that the size of the packed objects is insignificant in comparison to the total strip length. (In this context, see also [1].)

There are several other sources of difficulties of packing in two dimensions: the shape of the objects may be complicated (see [26] for an example from the clothing industry), or the domain of packing may be complicated. In this paper, we will deal with problems related to packing objects of simple shape (i.e., identical squares) into a *polygonal domain*: a connected region, possibly with holes, that has a boundary consisting of a total of n line segments, and the same number of vertices.

It should be noted that even when the structure of domain *and* objects are not complicated, only little is known – see the papers by Graham, Lubachevsky, and others [16, 19–21, 27–29] for packing identical disks into a strip, a square, a circle, or an equilateral triangle. Also, see Nelißen [34] for an overview of the so-called *pallet loading problem*, where we have to pack identical rectangles into a larger rectangle; it is still unclear whether this problem belongs to the class NP, since there may not be an optimal solution that can be described in polynomial time.

The following decision problem was shown to be NP-complete by Fowler et al. [15]; here and throughout the paper an *L-square* is a rectangle of size $L \times L$, and the number of vertices of a polygonal domain includes the vertices of all the holes it may have.

PACK(k, L):

Input: a polygonal domain P with n vertices, a parameter k, a parameter L.

Question: Can k many L-squares be packed into P?

PACK(k, L) is the decision problem for the following optimization problem:

$\max_k \text{PACK}(L)$:

Input: a polygonal domain P with n vertices

Task: Pack k many L-squares into P, such that k is as big as possible.

It was shown by Hochbaum and Maass [23] that $\max_k \text{PACK}(L)$ allows a polynomial time approximation scheme. The main contents of this paper is to examine several versions of the closely related problem

$\max_L \text{PACK}(k)$:

Input: a polygonal domain P with n vertices

Task: Pack k many $L \times L$ squares into P, such that L is as big as possible.

2 Preliminaries: Dispersion Problems

The problem $\max_L \text{PACK}(k)$ is a particular geometric *dispersion problem*. Problems of this type arise whenever the goal is to determine a set of positions, such that the objects are "far away" from each other. Examples for practical motivations are the location of oil storage tanks, ammunition dumps, nuclear power plants, hazardous waste sites – see the paper by Rosenkrantz, Tayi, and Ravi [36], who give a good overview, including the papers [6, 7, 9, 10, 18, 30, 32, 35, 39]. All these papers consider discrete sets of possible locations, so the problem can be considered as a generalized independent set problem in a graph. Special cases have been considered – see [5, 6]. However, for these discrete versions, the stated geometric difficulties do not come into play. In the following, we consider geometric versions, where the set of possible locations is given by a polygonal domain. We show the close connection to the packing problem and the polynomial approximation scheme by Hochbaum and Maass [23], but also a crucial difference: in general, if P$\neq$NP, it cannot be expected that the geometric dispersion problem can be approximated arbitrarily well.

When placing objects into a polygonal domain, we consider the following problem, where $d(v, w)$ is the geodesic distance between v and w:

$$\max_{S \subset P, |S|=k} \min_{v, w \in S} d(v, w).$$

This version corresponds to the dispersion problems in the discrete case and will be called *pure dispersion*.

In a geometric setting, we may not only have to deal with distances between locations; the distance of the dispersed locations to the boundary of the domain can also come into play. This yields the problem

$$\max_{S \subset P, |S|=k} \min_{v, w \in S} \{d(v, w), d(v, \partial P)\},$$

where ∂P denotes the boundary of the domain P. This version will be called *dispersion with boundaries*.

Finally, we may consider a generalization of the problem $\max_L \text{PACK}(k)$, which looks like a mixture of both previous variants:

$$\max_{S \subseteq P, |S|=k} \min_{v,w \in S} \{2d(v,w), d(v, \partial P)\}.$$

Since this corresponds to packing k many d-balls of maximum size into P, this variant is called *dispersional packing*.

It is also possible to consider other objective functions. Maximizing the average distance instead of the minimum distance can be shown to lead to a one-dimensional problem for pure dispersion (all points have to lie on the boundary of the convex hull of P). Details are omitted from this abstract.

Several distance functions can be considered for $d(v, w)$; the most natural ones are L_2 distances and L_1 or L_∞ distances. In the following, we concentrate on rectilinear, i.e., L_∞ distances. This means that we will consider packing squares with edges parallel to the coordinate axes. Most of the ideas carry over for L_2 distances by combining our ideas with the techniques by Hochbaum and Maass [23], and Fowler et al. [15]: again, it is possible to establish upper bounds on approximation factors, and get a factor $\frac{1}{2}$ by a simple greedy approach. Details are omitted from this abstract. We concentrate on the most interesting case of dispersion with boundaries, and only summarize the results for pure dispersion and dispersional packing; it is not hard to see that these variants are related via shrinking or expanding the domain P in an appropriate manner. See the full version of this paper or [2] for details.

The rest of this paper is organized as follows: In Section 3, we show that geometric dispersion with boundaries cannot be approximated arbitrarily well within polynomial time, unless P=NP; this result is valid, even if the polygonal domain has only axis-parallel edges, and distances are measured by the L_∞ metric. In Section 4, we give a strongly polynomial algorithm that approximates this case of geometric dispersion within a factor of $\frac{2}{3}$ of the optimum.

3 An upper bound on approximation factors

In this section, we give a sketch of an NP-completeness proof for geometric dispersion. Basically, we proceed along the lines of Fowler et al. [15], combined with the result by Lichtenstein [17, 25] about the NP-completeness of PLANAR 3SAT. We then argue that our proof implies an upper bound on approximation factors. In this abstract, we concentrate on the case of geometric dispersion with boundaries. In all figures, the boundaries correspond to the original boundaries of P, the interior is shaded in two colors. The lighter one corresponds to the part of the domain that is lost when shrinking P to accommodate for half of the considered distance $L^* = d(v, \partial P)$. The remaining dark region is the part that is feasible for packing $\frac{L^*}{2}$-squares.

Theorem 1. *Unless P=NP, there is no polynomial algorithm that finds a solution within more than $\frac{13}{14}$ of the optimum for rectilinear geometric dispersion with boundaries, even if the polygonal domain has only axis-parallel edges.*

Sketch: We give a reduction of PLANAR 3SAT. A 3SAT instance I is said to be an instance of PLANAR 3SAT, if the following bipartite graph G_I is planar: Every variable and every clause in I is represented by a vertex in G_I; two vertices are connected, if and only if one of them represents a variable that appears in the clause that is represented by the other vertex. See Figure 1 for an example.

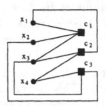

Fig. 1. The graph G_I representing the PLANAR 3SAT instance $(x_1 \lor x_2 \lor x_3) \land (\bar{x}_1 \lor \bar{x}_3 \lor x_4) \land (\bar{x}_2 \lor x_3 \lor \bar{x}_4)$

Proposition 2 (Lichtenstein) PLANAR 3SAT *is NP-complete.*

As a first step, we construct an appropriate planar layout of the graph G_I by using the methods of Duchet et al. [8], or Rosenstiehl and Tarjan [37]. Note that these algorithms produce layouts with all coordinates linear in the number of vertices of G_I.

Next, we proceed to represent variables, clauses, and connecting edges by suitable polygonal pieces. See Figure 2 for the construction of variable components.

Fig. 2. A variable component for dispersion with boundaries (top), a placement corresponding to "true" (center), and a placement corresponding to "false" (bottom)

The variable components are constructed in a way that allows basically two ways of dispersing a specific number of locations. One of them corresponds to a setting of "true", the other to a setting of "false". Depending on the truth setting, the adjacent connector components will have their squares pushed out or not. See Figure 5 (bottom) for the design of the connector components.

The construction of the clause components is shown in Figure 3: connector components from three variables (each with the appropriate truth setting) meet in such a way that there is a receptor region of "L" shape into which additional squares can be packed. Any literal that does not satisfy the clause forces one of the three corners of the L to be intersected by a square of the connector. Three additional squares can be packed if and only if one corner is not intersected, i.e., if the clause is satisfied.

From the above components, it is straightforward to compute the parameter k, the number of locations that are to be dispersed by a distance of 2. k is polynomial in the number of vertices of G_I and part of the input for the dispersion problem. All vertices of the resulting P have integer coordinates of small size, their number is polynomial in the number of vertices of G_I.

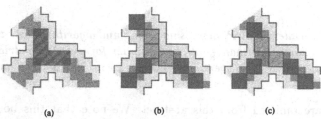

(a) (b) (c)

Fig. 3. A clause component for dispersion with boundaries and its receptor region (a), a placement corresponding to "true" (b), and a placement corresponding to "false" (c)

This shows that the problem is NP-hard. Now we need to argue that there cannot be a solution within more than $\frac{13}{14}$ of the optimum, if the PLANAR 3SAT instance cannot be satisfied.

Fig. 4. An upper bound on the approximation factor: variable components for 2-squares (left) and $(2 - 2\alpha)$-squares (right)

See Figure 4 (top) for the variable components. Indicated is a critical distance of $s = 11$. We show that it is impossible to pack an additional square into this section, even by locally changing the truth setting of a variable. Now suppose there was an approximation factor of $1 - \alpha$. This increases the feasible domain for packing squares by a width of 2α, and it decreases the size of these squares to $2 - 2\alpha$. If $\alpha < \frac{1}{s+3}$, then $\frac{s+1}{2}(2 - 2\alpha) > s + 2\alpha$, implying that it is impossible to place more than $\frac{s+1}{2}$ squares within the indicated part of the component. Similar arguments can be made for all other parts of the construction – see Figure 5. (Details are contained in the full version of the paper.) This shows that it is impossible to make local improvements in the packing to account for unsatisfied clauses, implying that we basically get the same solutions for value $2 - 2\alpha$ as for value 2, as long as $\alpha < \frac{1}{14}$. □

Along the same lines, we can show the following:

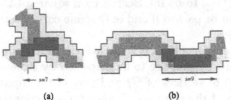

Fig. 5. An upper bound on the approximation factor: clause components (a) and connector components (b)

Theorem 3. *Unless P=NP, there is no polynomial algorithm that can guarantee a solution within more than $\frac{7}{8}$ of the optimum for pure geometric dispersion with L_∞ distances or for dispersional packing, even if the domain has only axis-parallel edges.*

Details are omitted from this abstract. We note that this bound can be lowered to $\frac{1}{2}$ if we do not require P to be a non-degenerate connected domain – see Figure 6 for the general idea. Further technical details are contained in the full version of the paper. It should be noted that the problem of covering a discrete point set, instead of "packing" into it, is well studied in the context of clustering – see the overview by Bern and Eppstein [4].

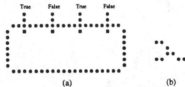

Fig. 6. An upper bound of $\frac{1}{2}$ on the approximation factor, if the domain P may be degenerate and disconnected – variable components (a) and clause components (b)

4 A $\frac{2}{3}$ approximation algorithm

In this section, we describe an approximation algorithm for geometric dispersion with axis-parallel boundaries in the case of L_∞ distances. We show that we can achieve an approximation factor of $\frac{2}{3}$. We use the following notation:

Definition 4 *The **horizontal α-neighborhood** of a d square Q is a rectangle of size $((d+\alpha) \times d)$ with the same center as Q.*
For a polygonal domain P and a distance r, $\boldsymbol{P-r}$ is the polygonal domain $\{p \in P \mid d(p, \partial P) \geq r\}$, obtained by shrinking P. Similarly, $\boldsymbol{P+r}$ is the domain $\{p \in \mathbb{R}^2 \mid d(p, P) \leq r\}$, obtained by expanding P. Note that $P+r$ is a polygonal domain for rectilinear distances.
$\boldsymbol{Par(P)} := \{(e_i, e_j) \mid e_i \| e_j ;\ e_i, e_j \in E(P)\}$ is the set of all pairs of parallel edges of P. $\boldsymbol{Dist(e_i, e_j)}$ (for $(e_i, e_j) \in Par(P)$) is the distance of the edges e_i and e_j.
With $\boldsymbol{AS(P, d, l)}$, we call the approximation scheme by Hochbaum and Maass

for $\max_k \text{PACK}(L)$, *where P is the feasible domain, d is the size of the squares, and l is the width of the strips, guaranteeing that the number of packed squares is at least within a factor of* $\left(\frac{l-1}{l}\right)^2$ *of the optimum.*

Note that the approximation scheme $AS(P, d, l)$ can be modified for axis-parallel boundaries, such that the resulting algorithms are strongly polynomial: If the number of squares that can be packed is not polynomial in the number n of vertices of P, then there must be two "long" parallel edges. These can be shortened by cutting out a "large" rectangle, which can be dealt with easily. This procedure can be repeated until all edges are of length polynomially bounded in n. (Details are contained in the full version of the paper. Also, see [2, 40].)

The idea of the algorithm is the following: Use binary search over the size d of the squares in combination with the approximation scheme by Hochbaum and Maas for $\max_k \text{PACK}(L)$ in order to find the largest size d of squares where the approximation scheme guarantees a packing of k many d-squares into the domain $P - \frac{d}{2}$, with the optimum number of d-squares guaranteed to be strictly less than $\frac{3k}{2}$. Then the following crucial lemma guarantees that it is impossible to pack k squares of size at least $\frac{3d}{2}$ into $P - \frac{3d}{4}$, implying a $\frac{2}{3}$ approximation algorithm.

Lemma 5 *Let P be a polygonal domain, such that k many $\frac{3d}{2}$-squares can be packed into $P - \frac{3d}{4}$. Then at least $\frac{3}{2}k$ many d-squares can be packed into $P - d/2$.*

Proof. Consider a packing of k many $\frac{3d}{2}$-squares into $P - \frac{3d}{4}$.
Clearly, we have:

$$(P - \frac{3d}{4}) + \frac{d}{4} \subseteq P - d/2. \tag{1}$$

For constructing a packing of d-squares, it suffices to consider the domain that is covered by the $\frac{3d}{2}$-squares instead of $P - \frac{3d}{4}$. After expanding this domain by $\frac{d}{4}$, we get a subset of $P - d/2$ by (1). In the following, we construct a packing of d-squares. At any stage, the following Observation 6 is valid.

Observation 6 *Suppose the feasible space for packing d-squares contains the horizontal $\frac{d}{4}$-neighborhoods of a set of disjoint $\frac{3d}{2}$-squares. Then there exists a $\frac{3d}{2}$-square Q that has left-most position among all remaining squares, i. e., to the left of Q, the horizontal $\frac{d}{4}$-neighborhood of Q does not overlap the horizontal $\frac{d}{4}$-neighborhood of any other $\frac{3d}{2}$-square.*

While there are $\frac{3d}{2}$-squares left, consider a leftmost $\frac{3d}{2}$-square Q. We distinguish cases, depending on the relative position of Q with respect to other remaining $\frac{3d}{2}$-squares. See Figure 7.

Details are omitted from this abstract for lack of space. At any stage, a set of one, two, or three $\frac{3}{2}d$-squares that includes a leftmost $\frac{3}{2}d$-square is replaced by a set of two, three, or five d-squares This iteration is performed while there are $\frac{3}{2}d$-squares left. It follows that we can pack at least $\frac{3}{2}k$ many d-squares into $P - \frac{d}{2}$. □

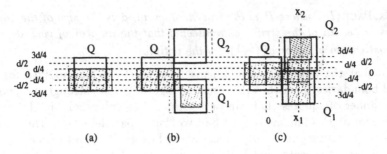

Fig. 7. Constructing a packing of d-squares

In the following, we give a formal description of the binary search algorithm, and argue why it is possible to terminate the binary search in polynomial time.

Algorithm 7

Input: *polygonal domain P, positive integer k.*

Output: $A_{Dis}(P,k) := d$ *is the minimum L_∞ distance between a location and the boundary or between two locations.*

1. **For all** $(e_i, e_j) \in Par(P)$ **do**
2. **While** d_{ij} *undetermined, perform binary search as follows:*
   - (a) $s_{max} := k + 1$ and $s_{min} := 2$ and $d_{max} := \frac{2}{3}Dist(e_i, e_j)/(s_{max})$ and $d_{min} := \frac{2}{3}Dist(e_i, e_j)/(s_{min})$.
   - (b) **If** $AS(P - d_{max}/2, d_{max}, 6) < k$ **then** $d_{ij} = 0$.
   - (c) **If** $AS(P - d_{min}/2, d_{min}, 6) \geq k$ **then** $d_{ij} := \frac{2}{3}Dist(e_i, e_j)$.
   - (d) **While** $s_{max} > s_{min} + 1$ **do**
   - (e) $s := \lceil (s_{max} + s_{min})/2 \rceil$ and $d := \frac{2}{3}Dist(e_i, e_j)/s$.
   - (f)   **If** $AS(P - d/2, d, 6) \geq k$ **then** $s_{max} := s$.
   - (g)   **Else** $s_{min} := s$.
   - (h) $d_{ij} := d$.
3. **Output** $d := \max\{d_{ij} \mid (e_i, e_j) \in Par(P)\}$ **and exit.**

Theorem 8. *For rectilinear geometric dispersion with boundaries of k locations in a polygonal domain P with axis-parallel boundaries and n vertices, Algorithm 7 computes a solution $A_{Dis}(P, k)$, such that*

$$A_{Dis}(P, k) \geq \frac{2}{3}OPT(P, k).$$

The running time is strongly polynomial.

Proof. It is not hard to see that there are only finitely many values for the optimal value between the k points. More precisely, we can show that for the optimal distance d_{opt}, the following holds:

There is a pair of edges $(e_i, e_j) \in Par(P)$, such that

$$d_{opt} = \frac{Dist(e_i, e_j)}{s_{ij}} \qquad \text{for some } 2 \leq s_{ij} \leq k + 1. \tag{2}$$

In order to determine an optimal solution, we only need to consider values that satisfy Equation (2). For every pair of parallel edges of P, there are only k possible values for an optimal distance of points. Thus, there can be at most $O(n^2 k)$ many values that need to be considered.

We proceed to show that the algorithm guarantees an approximation factor of $\frac{2}{3}$.

By binary search, the algorithm determines for every pair of edges $(e_i, e_j) \in Par(P)$ of P a d_{ij} with the following properties:

1. $\frac{3}{2} d_{ij} = Dist(e_i, e_j)/s_{ij}$ $(2 \leq s_{ij} \leq k+1)$ is a possible optimal value for the distance of k points that have to be dispersed in P.
2. Using the approximation scheme [23], at least k many d_{ij}-squares can be packed into $P - d_{ij}/2$, with $d_{ij} = Dist(e_i, e_j)/(s_{ij})$.
3. If $s_{ij} > 2$, then for $\tilde{d}_{ij} := \frac{2}{3} Dist(e_i, e_j)/(s_{ij} - 1)$, we cannot pack k many $\tilde{d}_{ij}$-squares into $P - \tilde{d}_{ij}/2$ with the help of the approximation scheme.

Property 1 follows from (2), Properties 2 and 3 hold as a result of the binary search.

From Lemma 5, we know that at least $\frac{3}{2} k$ many $\frac{2}{3} d_{opt}$-squares can be packed into $P - \frac{1}{3} d_{opt}$, since k many d_{opt}-squares can be packed into $P - d_{opt}/2$.

Let $k_{opt}(P - \frac{1}{3} d_{opt}, \frac{2}{3} d_{opt})$ be the optimal number of $\frac{2}{3} d_{opt}$-squares that can be packed into $P - \frac{1}{3} d_{opt}$. With the parameter $l = 6$, the approximation scheme [23] guarantees an approximation factor of $(\frac{5}{6})^2$. This implies:

$$k_{opt}\left(P - \frac{1}{3} d_{opt}, \frac{2}{3} d_{opt}\right) \leq \left(\frac{6}{5}\right)^2 AS\left(P - \frac{1}{3} d_{opt}, \frac{2}{3} d_{opt}, 6\right)$$

$$< \frac{3}{2} AS\left(P - \frac{1}{3} d_{opt}, \frac{2}{3} d_{opt}, 6\right).$$

It follows that

$$\frac{3}{2} AS\left(P - \frac{1}{3} d_{opt}, \frac{2}{3} d_{opt}, 6\right) > k_{opt}\left(P - \frac{1}{3} d_{opt}, \frac{2}{3} d_{opt}\right) \geq \frac{3}{2} k$$

This means that at least k squares are packed when the approximation scheme is called with a value of at most $\frac{2}{3} d_{opt}$.

For $\tilde{d}_{ij}$ this means that $\tilde{d}_{ij} > \frac{2}{3} d_{opt}$ and therefore $\frac{3}{2} \tilde{d}_{ij} = Dist(e_i, e_j)/(s_{d_{ij}} + 1) > d_{opt}$.

Hence, for every pair $(e_i, e_j) \in Par(P)$ of edges, the algorithm determines a value d_{ij} that satisfies $\frac{3}{2} d_{ij} = Dist(e_i, e_j)/s_{d_{ij}}$ and is a potential optimal value, and the next larger potential value is strictly larger than the optimal value.

The algorithm returns the d with $d = \max\{d_{ij} \,|\, (e_i, e_j) \in Par(P)\}$. Therefore, $\frac{3}{2} d \geq d_{opt}$, implying

$$A_{Dis}(P, k) = d \geq \frac{2}{3} d_{opt} = \frac{2}{3} OPT(P, k),$$

proving the approximation factor.

The total running time is $O(\log k \cdot n^{40})$. Note that the strongly polynomial modified version of the approximation scheme [23] takes $O(l^2 \cdot n^2 \cdot n^{l^2})$, i.e., $O(n^{38})$ with $l = 6$.

For the case of general polygonal domains, where the boundaries of the domain are not necessarily axis-parallel, Lemma 5 is still valid. In the full version of the paper, we discuss approximation for this more general case.

Without further details, we mention

Theorem 9. *Pure geometric dispersion and dispersional packing can be approximated within a factor of $\frac{1}{2}$ in (strongly) polynomial time.*

These factors can be achieved without use of the approximation scheme via a straightforward greedy strategy; the approximation factor follows from the fact that any packing of k $2d$-balls guarantees a packing of $2k$ many d-balls, and a greedy packing guarantees a $\frac{1}{2}$-approximation for $\max_k \mathrm{PACK}(L)$.

5 Conclusions

We have presented upper and lower bounds for approximating geometric dispersion problems. In the most interesting case of a non-degenerate, connected domain, these bounds still leave a gap; we believe that the upper bounds can be improved. It would be very interesting if some of the lower bounds of $\frac{1}{2}$ could be improved. If we assume that the area of P is large, it is not very hard to see that an optimal solution can be approximated much better. It should be possible to give some quantification along the lines of an asymptotic polynomial time approximation scheme.

It is clear from our above results that similar upper and lower bounds can be established for L_2 distances.

Like for packing problems, there are many possible variants and extensions. One of the interesting special cases arises from considering a *simple polygon*, i. e., a polygonal region without holes. The complexity of this problem is unknown, even if the simple polygon is *rectilinear*, i. e., all its edges are axis-parallel.

Conjecture 10 *The problem* $\mathrm{PACK}(k, L)$ *for* L_∞ *distances is polynomial for the class of simple polygons P.*

Acknowledgments

The second author would like to thank Joe Mitchell, Estie Arkin, and Steve Skiena for discussions at the Stony Brook Computational Geometry Problem Seminar [33], which gave some initial motivation for this research.

We thank Jörg Schepers, Joe Mitchell, and the anonymous referees for helpful comments.

References

1. B. S. Baker, D. J. Brown and H. K. Katseff. A 5/4 algorithm for two-dimensional packing, *Journal of Algorithms*, **2**, 1981, 348–368.
2. Christoph Baur. Packungs- und Dispersionsprobleme. Diplomarbeit, Mathematisches Institut, Universität zu Köln, 1997.
3. J. E. Beasley. An exact two–dimensional non-guillotine cutting tree search procedure, *Operations Research*, **33**, 1985, 49–64.
4. M. Bern and D. Eppstein. Clustering. Section 8.5 of the chapter *Approximation algorithms for geometric problems*, in: D. Hochbaum (ed.): Approximation Algorithms for NP-hard Problems. PWS Publishing, 1996.
5. B. K. Bhattacharya and M. E. Houle. Generalized maximum independent set for trees. To appear in: *Journal of Graph Algorithms and Applications*, 1997. mike@cs.newcastle.edu.au.
6. R. Chandrasekaran and A. Daughety. Location on tree networks: p-centre and n-dispersion problems. *Mathematics of Operations Research*, **6**, 1981, 50–57.
7. R. L. Church and R. S. Garfinkel. Locating an obnoxious facility on a network. *Transportation Science*, **12**, 1978, 107–118.
8. P. Duchet, Y. Hamidoune, M. Las Vergnas, and H. Meyniel. Representing a planar graph by vertical lines joining different levels. *Discrete Mathematics*, **46**, 1983, 319–321.
9. E. Erkut. The discrete p–dispersion problem. *European Journal of Operational Research*, **46**, 1990, 48–60.
10. E. Erkut and S. Neumann. Comparison of four models for dispersing facilities. *European Journal of Operations Research*, **40**, 1989, 275–291.
11. S. P. Fekete and J. Schepers. A new exact algorithm for general orthogonal d-dimensional knapsack problems. *Algorithms – ESA '97*, Springer Lecture Notes in Computer Science, vol. 1284, 1997, 144–156.
12. S. P. Fekete and J. Schepers. On more-dimensional packing I: Modeling. Technical report, ZPR 97-288. Available at http://www.zpr.uni-koeln.de/~paper
13. S. P. Fekete and J. Schepers. On more-dimensional packing II: Bounds. Technical report, ZPR 97-289. Available at http://www.zpr.uni-koeln.de/~paper
14. S. P. Fekete and J. Schepers. On more-dimensional packing III: Exact Algorithms. Technical report, ZPR 97-290. Available at http://www.zpr.uni-koeln.de/~paper
15. R. J. Fowler, M. S. Paterson, and S. L. Tanimoto. Optimal packing and covering in the plane are NP–complete. *Information Processing Letters*, **12**, 1981, 133–137.
16. Z. Füredi. The densest packing of equal circles into a parallel strip. *Discrete & Computational Geometry*, 1991, 95–106.
17. M. R. Garey and D. S. Johnson. *Computers and Intractability: A Guide to the theory of NP–Completeness*. W. H. Freeman and Company, San Francisco, 1979.
18. A. J. Goldmann and P. M. Dearing. Concepts of optimal locations for partially noxious facilities. *Bulletin of the Operational Research Society of America*. **23**, B85, 1975.
19. R. L. Graham and B. D. Lubachevsky. Dense packings of equal disks in an equilateral triangle: from 22 to 34 and beyond. *The Electronic Journal of Combinatorics*, **2**, 1995, #A1.
20. R. L. Graham and B. D. Lubachevsky. Repeated patterns of dense packings of equal disks in a square. *The Electronic Journal of Combinatorics* **3**, 1996, #R16.

21. R. L. Graham, B. D. Lubachevsky, K. J. Nurmela, and P. R. J. Östergård. Dense packings of congruent circles in a circle. Manuscript, 1996. bdl@research.att.com.

22. E. Hadjiconstantinou and N. Christofides. An exact algorithm for general, orthogonal, two-dimensional knapsack problems, *European Journal of Operations Research*, **83**, 1995, 39-56.

23. D. S. Hochbaum and W. Maass. Approximation schemes for covering and packing problems in image processing and VLSI. *Journal of the ACM*, **32**, 1985, 130-136.

24. C. Kenyon and E. Remila, Approximate strip packing. *Proc. of the 37th Annual Symposium on Foundations of Computer Science (FOCS 96)*, 142-154.

25. D. Lichtenstein. Planar formulae and their uses. *SIAM Journal on Computing*, 11, 1982, 329-343.

26. Z. Li and V. Milenkovic. A compaction algorithm for non-convex polygons and its application. *Proc. of the Ninth Annual Symposium on Computational Geometry*, 1993, 153-162.

27. B. D. Lubachevsky and R. L. Graham. Curved hexagonal packings of equal disks in a circle. Manuscript. bdl@research.att.com.

28. B. D. Lubachevsky, R. L. Graham, and F. H. Stillinger. Patterns and structures in disk packings. 3rd Geometry Festival, Budapest, Hungary, 1996. Manuscript. bdl@research.att.com.

29. C. D. Maranas, C. A. Floudas, and P. M. Pardalos. New results in the packing of equal circles in a square. *Discrete Mathematics* **142**, 1995, 287-293.

30. M. V. Marathe, H. Breu, H. B. Hunt III, S. S. Ravi, and D. J. Rosenkrantz. Simple heuristics for unit disk graphs. *Networks*, **25**, 1995, 59-68.

31. S. Martello and P. Toth, *Knapsack Problems - Algorithms and Computer Implementations,* Wiley, Chichester, 1990.

32. I. D. Moon and A. J. Goldman. Tree location problems with minimum separation. *Transactions of the Institute of Industrial Engineers*, **21**, 1989, 230-240.

33. J. S. B. Mitchell, Y. Lin, E. Arkin, and S. Skiena. *Stony Brook Computational Geometry Problem Seminar.* Manuscript. jsbm@ams.sunysb.edu.

34. J. Nelißen. *New approaches to the pallet loading problem.* Technical Report, RWTH Aachen, Lehrstuhl für Angewandte Mathematik, 1993.

35. S. S. Ravi, D. J. Rosenkrantz, and G. K. Tayi. Heuristic and special case algorithms for dispersion problems. *Operations Research*, **42**, 1994, 299-310.

36. D. J. Rosenkrantz, G. K. Tayi, and S. S. Ravi. Capacitated facility dispersion problems. Manuscript, submitted for publication, 1997. ravi@cs.albany.edu.

37. P. Rosenstiehl and R. E. Tarjan. Rectilinear planar layouts and bipolar orientations of planar graphs. *Discrete & Computational Geometry*, 1, 1986, 343-353.

38. J. Schepers. Exakte Algorithmen für orthogonale Packungsprobleme. Dissertation, Mathematisches Institut, Universität zu Köln, 1997.

39. A. Tamir. Obnoxious facility location on graphs. *SIAM Journal on Discrete Mathematics*, 4, 1991, 550-567.

40. M. Wottawa. Struktur und algorithmische Behandlung von praxisorientierten dreidimensionalen Packungsproblemen. Dissertation, Mathematisches Institut, Universität zu Köln, 1996.

Approximating k-outconnected Subgraph Problems

Joseph Cheriyan[1], Tibor Jordán[2], and Zeev Nutov[1]

[1] Department of Combinatorics and Optimization, University of Waterloo, Waterloo, Ontario, Canada N2L 3G1,
jcheriyan, znutov@math.uwaterloo.ca
[2] Institut for Matematik og Datalogi, Odense Universitet, Odense, Denmark,
tibor@imada.ou.dk

Abstract. We present approximation algorithms and structural results for problems in network design. We give improved approximation algorithms for finding min-cost k-outconnected graphs with either a single root or multiple roots for (i) uniform costs, and (ii) metric costs. The improvements are obtained by focusing on single-root k-outconnected graphs and proving (i) a version of Mader's critical cycle theorem and (ii) an extension of a splitting off theorem by Bienstock et al.

1 Introduction

We study some NP-hard problems from the area of network design. We are interested in approximation algorithms as well as structural results, and we use new structural results to obtain improved approximation guarantees.

A basic problem in network design is to find a min-cost k-connected spanning subgraph. (In this paper k-connectivity refers to k-node connectivity.) The problem is NP-hard, and there is an $O(\log k)$-approximation algorithm due to [13]. A generalization is to find a min-cost subgraph that has at least k_{vw} openly disjoint paths between every node pair v, w, where $[k_{vw}]$ is a prespecified "connectivity requirements" matrix. No poly-logarithmic approximation algorithm for this problem is known. For the special case when each k_{vw} is in $\{0, 1, 2\}$, [13] gives a 3-approximation algorithm. The min-cost k-outconnected subgraph problem is "sandwiched" between the basic problem and the general problem. There are two versions of the problem. In the single-root version, there is a root node r, and the connectivity requirement is $k_{rv} = k$, for all nodes v ($k_{vw} = 0$ if $v \neq r$ and $w \neq r$). This problem is NP-hard, even for $k = 2$ and uniform costs. (To see this, note that a 2-outconnected subgraph of a graph G has $\leq |V(G)|$ edges iff G has a Hamiltonian cycle.) In the multi-root version, there are $q \geq 1$ root nodes $r_1, \ldots, r_q$, and the connectivity requirement is $k_{r_i v} = k_i$ for $i = 1, \ldots, q$ and for all nodes v; note that the connectivity requirements $k_1, \ldots, k_q$ of the roots $r_1, \ldots, r_q$ may be distinct. Consider a directed variant of the single-root problem: given a digraph with a specified root node r, find a min-cost subdigraph that contains k openly disjoint $r \rightarrow v$ directed paths for all nodes v. A polynomial-time algorithm for finding an optimal solution to this digraph problem is given in [4]. Based on this, [9] gives a 2-approximation algorithm for the undirected single-root problem. For the multi-root problem, a $2q$-approximation algorithm follows by sequentially applying the 2-approximation algorithm of [9] to each of the roots $r_1, \ldots, r_q$. To the best of our knowledge, no better approximation guarantees were known even for uniform costs (i.e., the min-size k-outconnected subgraph problem) and for metric costs (i.e., edge costs satisfying the triangle inequality). There has been extensive recent research on approximation algorithms for NP-hard network design problems with uniform costs and with metric costs, see the survey in [8] and see [1, 3, 5, 9], etc. For the uniform cost multi-root problem, we improve the approximation guarantee from $2q$ to $\min\{2, \frac{k+2q-1}{k}\}$ (this implies a $1 + \frac{1}{k}$ approximation algorithm for the uniform cost single-root problem), and for the metric cost multi-root problem, we improve the approximation guarantee from $2q$ to 4 (in fact, to $3 + \frac{k_s}{k_m}$, where k_m and k_s are the largest and the second largest requirements, respectively).

The multi-root problem appears to have been introduced in [12], and for some special cases with $\max\{k_i \mid i = 1, \ldots, q\} \leq 3$, approximation guarantees better than $2q$ are derived in [12]. An application of the multi-root problem to "mobile robot flow networks" is described in [12]. We skip the applications in this paper.

Our results on min-size k-outconnected subgraphs are based on a new structural result that extends a result of Mader [10]. Mader's theorem (see Theorem 2) states that in a k-node connected graph, a cycle of critical edges must be incident to a node of degree k. Our result is similar to Mader's statement, except it applies to single-root k-outconnected graphs, and "critical" edge means critical for k-outconnectivity, see Theorem 1. Our proof is similar to Mader's proof but as far as we can see, neither of the two results generalizes the other.

The results on the metric version of the min-cost k-outconnected subgraph problem are based on a partial extension of a splitting off theorem due to Bienstock et al. [1]. They proved that if the edges incident to a vertex r are all critical with respect to k-connectivity in a k-connected graph G ($k \geq 2$) and the degree $d(r)$ is at least $k + 2$ then either there exists a pair of edges incident to r which can be split off preserving k-connectivity or there exist two pairs, one of them is incident to r, such that splitting off both preserves k-connectivity. We prove the corresponding statement in the case when G is k-outconnected from r, the edges incident to r are critical with respect to k-outconnectivity and $d(r) \geq 2k + 2$. It turns out that our result implies the splitting result of [1] when $d(r) \geq 2k + 2$.

Definitions and notation

We consider only *simple* graphs, that is, the graphs have no multi-edges and no loops. Throughout the paper, when discussing a problem, we use $G = (V, E)$ to denote the input graph, and *opt* to denote the optimal value of the problem on G; also, we assume that G has a feasible solution to the problem. The number of nodes of G is denoted by n, so $n = |V|$. *Splitting off* two edges su, sv means deleting su and sv and adding a new edge uv. For a set $X \subseteq V$ of nodes $\Gamma(X) := \{y \in V - X : xy \in E \text{ for some } x \in X\}$ denotes its set of *neighbours*.

A graph is said to be *k-outconnected from* a node r, if for every node v, there exist k openly disjoint $v \leftrightarrow r$ paths. The node r is called the *root*. Clearly, for a fixed root node r, a k-node connected graph is k-outconnected from r (by Menger's theorem), but the reverse is not true. For example, we can take several different k-node connected graphs $G_1, G_2, \ldots$, choose an arbitrary node w_i in each G_i, and identify $w_1, w_2, \ldots$ into a single node r. The resulting graph has node connectivity one, but is k-outconnected from r. It can be seen that if G is k-outconnected from r, then every separator (node cut) of cardinality $< k$ must contain the root r. Consequently, if G is not k-node connected, then there are two neighbours v, w of r such that G has at most $(k - 1)$ openly disjoint $v \leftrightarrow w$ paths.

2 Structural results for *k*-outconnected subgraphs

2.1 Reducible root sequences

First we show a simple but useful observation which shows that there is no loss of generality in assuming that the number of roots for the multi-root outconnected subgraph problem is at most k, where k is the maximum connectivity requirement; moreover, if the number of roots is k, then each root r_i has $k_i = k$. Suppose that we are given a set $R = \{r_1, \ldots, r_q\}$ of root nodes together with their connectivity requirements $\boldsymbol{k} = (k_1, \ldots, k_q)$. Without loss of generality we may assume that $k_1 \leq \ldots \leq k_q = k$.

Lemma 1. *If there is an index $j = 1, \ldots, q$ such that $k_j \leq q - j$, then a graph is $\boldsymbol{k}$-outconnected from R iff it is $\boldsymbol{k'}$-outconnected from R', where $\boldsymbol{k'} = (k_{j+1}, \ldots, k_q)$ and $R' = \{r_{j+1}, \ldots, r_q\}$.* □

Corollary 1. *If $q > k$, then we can replace R by $R - \{v_1, \ldots, v_{q-k}\}$ and we can change $\boldsymbol{k}$ appropriately. If $q = k$ and say $k_1 < k$, then we can replace R by $R - \{v_1\}$ and we can change $\boldsymbol{k}$ appropriately.* □

2.2 The extension of Mader's theorem

This section has a proof of the following result.

Theorem 1. *Let H be a graph that is k-outconnected from a node r. In H, a cycle consisting of critical edges must be incident to a node $v \neq r$ such that $\deg(v) = k$. (Here, critical is with respect to k-outconnectivity, i.e., an edge e is called critical if $H - e$ is not k-outconnected from r.)*

For comparison, we state Mader's theorem.

Theorem 2 (Mader [10]). *In a k-node connected graph H', a cycle of critical edges must be incident to a node of degree k. (Here, critical is with respect to k-node connectivity, i.e., an edge e is called critical if $H' - e$ is not k-node connected.)* □

Remark: Theorem 1 does not seem to have any obvious extension to multi-root outconnected graphs. Here is an example H with two roots r_1, r_2 and $k = 3$ that is k-outconnected from each of r_1 and r_2 such that there is a cycle of critical edges such that each incident node has degree $\geq 4 > k = 3$. Take a 6-cycle $v_1, \ldots, v_6, v_1$ and add two nodes v_7 and v_8, and add the following edges incident to v_7 or v_8: $v_7v_1, v_7v_2, v_7v_5, v_7v_6, v_8v_2, v_8v_3, v_8v_4, v_8v_5$. Let the roots be $r_1 = v_2$, $r_2 = v_5$. Note that each edge of H is critical, either for 3-outconnectivity from r_1 or for 3-outconnectivity from r_2. The cycle $C = v_7, r_1, v_8, r_2, v_7$ has the property stated above. We have examples where the cycle in question is incident to *no* root.

The following corollary of Theorem 1 gives an extension of Theorem 2.

Corollary 2. *Let G be a k-node connected graph, and let C be a cycle of critical edges that is incident to exactly one node v_0 with $\deg(v_0) = k$ (so $\deg(v) \geq k+1$ for all nodes $v \in V(C) - \{v_0\}$). Then there exists an edge e in C such that every $(k-1)$-separator S of $G - e$ contains v_0.* □

Our proof of Theorem 1 is based on two lemmas. The second of these is similar to the key lemma used by Mader in his proof of Theorem 2; we skip the full proof, and instead refer the interested reader to [10, Lemma 1] or to [2, Lemma 4.4].

Lemma 2. *Let $H = (V, E)$ be k-outconnected from r. Let v be a neighbour of r with $\deg(v) \geq k + 1$, and let $vw \neq vr$ be an edge. Then in $H - vw$, there are k openly disjoint $r \leftrightarrow v$ paths.* □

Corollary 3. *Let $H = (V, E)$ be k-outconnected from r, and let $e = vw$ be a critical edge (with respect to k-outconnectivity from r). Then in $H - e$ there is a $(k-1)$-separator $S_e \subset V - \{r, v, w\}$ such that $H - e - S_e$ has exactly two components, one containing v and the other containing w.* □

For a critical edge $e = vw$ of H, let S_e denote a $(k-1)$-separator as in the corollary, and let the node sets of the two components of $H - e - S_e$ be denoted by $D_{v,e}$ and $D_{w,e}$, where $v \in D_{v,e}$ and $w \in D_{w,e}$.

Lemma 3 (Mader). *Let $H = (V, E)$ be k-outconnected from r. Let $v \neq r$ be a node with $\deg(v) \geq (k+1)$, and let $e = vw$ and $f = vx$ be two critical edges. Let $S_e, D_{v,e}, D_{w,e}$ and $S_f, D_{v,f}, D_{x,f}$ be as defined above.*

(1) Then $D_{w,e}$ and $D_{x,f}$ have no nodes in common, i.e., $D_{w,e} \cap D_{x,f} = \emptyset$.
(2) If $r \in D_{w,e}$, then $r \in D_{v,f}$, and symmetrically, if $r \in D_{x,f}$, then $r \in D_{v,e}$.

Proof. For the proof of part (1), we refer the reader to Mader's proof, see [10, Lemma 1] or [2, Lemma 4.4].

For part (2), note that

$$D_{w,e} = (D_{w,e} \cap D_{v,f}) \cup (D_{w,e} \cap S_f) \cup (D_{w,e} \cap D_{x,f}).$$

If r is in $D_{w,e}$, then r is in $D_{w,e} \cap D_{v,f}$, because $r \notin S_f$ (by hypothesis), and $r \notin D_{w,e} \cap D_{x,f}$ (by part (1)). □

Proof. (of Theorem 1) The proof is by contradiction. Let $C = v_0, v_1, v_2, \ldots, v_p, v_0$ be a cycle of critical edges, and let every node in $V(C) - \{r\}$ have degree $\geq k+1$. The case when C is not incident to r is easy to handle. There is a reduction to k-node connected graphs, and the contradiction comes from Mader's theorem. Here is the reduction: replace the root r by a clique Q on $\deg(r)$ nodes and replace each edge incident to r by an edge incident to a distinct node of Q.

Claim 1: If H is k-outconnected from r, then the new graph is k-node connected. Moreover, for every edge $vw \in E(H)$, if vw is critical in H (with respect to k-outconnectivity from r), then vw is critical in the new graph with respect to k-node connectivity.

From Claim 1 and Mader's theorem (Theorem 2), we see that there is a contradiction if C is not incident to r.

Now, suppose that C is incident to r. Recall that $C = v_0, v_1, v_2, \ldots, v_p, v_0$, and let $r = v_0$. For each edge $v_i v_{i+1}$ in C (indexing modulo p), let us revise our notation to $S_i = S_{v_i v_{i+1}}$, $D'_i = D_{v_i, v_i v_{i+1}}$, $D''_i = D_{v_{i+1}, v_i v_{i+1}}$. (So $S_i \subseteq V - \{r, v_i, v_{i+1}\}$ has cardinality $k-1$, and $G - v_i v_{i+1} - S_i$ has two components with node sets D'_i and D''_i, where v_i is in the former and v_{i+1} is in the latter.)

The next claim follows easily by induction on i, using Lemma 3, part (2); the induction basis is immediate.

Claim 2: For each $i = 0, 1, 2, 3, \ldots, p$, the root r is in D'_i.

Claim 2 gives the contradiction needed to prove the theorem, because the claim states that r is in the component of v_p (rather than the component of $v_0 = r$) in $H - v_p v_0 - S_p$. $\square$

2.3 Splitting off edges from the root in a k-outconnected graph

In this subsection we present our result on the existence of pairs of edges which can be split off from the root r. First let us recall the following result of Bienstock et al.

Theorem 3 (Bienstock et al. [1]). *Let $G = (V, E)$ be a k-connected graph, $|V| \geq 2k$, $k \geq 2$. Suppose that the edges incident to a node $r \in V$ are all critical with respect to k-connectivity and $d(r) \geq k + 2$. Then either there exists a pair ru, rv of edges incident to r which can be split off preserving k-connectivity or for any pair ru, rv there exists a pair sw, sz such that splitting off both pairs preserves k-connectivity.* $\square$

We prove the following.

Theorem 4. *Let $G' = (V + r, E')$ be a graph that is k-outconnected from r and where the edges incident to r are all critical with respect to k-outconnectivity and $d(r) \geq 2k + 1$. Then either there exists a pair of edges incident to r which can be split off preserving k-outconnectivity or G' is k-node connected.*

Proof. Let $G' = (V + r, E')$ be a simple undirected graph with a designated *root* node r. For each $v \in V$ let $f(v) := 1$ if $rv \in E'$ and $f(v) := 0$ otherwise. For some $\emptyset \neq X \subseteq V$ let $f(X) = \sum_{v \in X} f(v)$. Let $g(X) := |\Gamma_{G'-r}(X)| + f(X)$ be also defined on the non-empty subsets of V, where $\Gamma_{G'-r}(X)$ is the set of neighbours of X in the graph $G' - r$.

Lemma 4. *G' is k-outconnected from r if and only if*

$$g(X) \geq k \quad \text{for every} \quad \emptyset \neq X \subseteq V. \tag{1}$$

Proof. The lemma follows easily from Menger's theorem. $\square$

Note that the function $g(X)$ is *submodular* (since it is obtained as the sum of a submodular and a modular function). That is, $g(X) + g(Y) \geq g(X \cap Y) + g(X \cup Y)$ for every $X, Y \subseteq V$.

In what follows assume that $G' = (V + r, E')$ is k-outconnected from the root r ($k \geq 2$) and every edge rv incident to r in G' is critical with respect to k-outconnectivity (that is, $G - rv$ is no longer k-outconnected). It will be convenient to work with the graph $G = (V, E) := G' - r$ and the functions f, g defined above. By Lemma 4 a pair ru, rv of edges is *admissible* for splitting in G' (that is, splitting off the pair preserves k-outconnectivity) if and only if decreasing $f(u)$ and $f(v)$ by one and adding a new edge uv in G preserves (1). Therefore such a pair u, v of nodes in G (with $f(u) = f(v) = 1$) is also called *admissible*. Otherwise the pair u, v is *illegal*. Thus we may consider a graph $G = (V, E)$ with a function $f : V \rightarrow \{0, 1\}$ for which G satisfies (1) and for which f is minimal with respect to this property (that is, decreasing $f(v)$ by one for any $v \in V$ with $f(v) = 1$ destroys (1)), and search for admissible pairs of nodes. A node $u \in V$ with $f(u) = 1$ is called *positive*. Let F denote the set of positive nodes in G.

From now on suppose that each pair $x, y \in F$ is illegal in G. It is not difficult to see that a pair x, y is illegal if and only if one of the following holds:

(i) there exists an $X \subseteq V$ with $x, y \in X$ and $g(X) \leq k + 1$,

(iia) there exists an $X \subseteq V$ with $x \in X$, $y \in \Gamma(X)$ and $g(X) = k$,

(iib) there exists an $X \subseteq V$ with $y \in X$, $x \in \Gamma(X)$ and $g(X) = k$.

A set $X \subseteq V$ with $g(X) \leq k + 1$ is called *dangerous*. If $g(X) = k$ then X is *critical*. The minimality of f implies that for every positive node x there exists a critical set $X \subseteq V$ with $x \in X$. Using that g is submodular, standard arguments give the following:

Lemma 5. *(1) The intersection and union of two intersecting critical sets are both critical,*

(2) for two intersecting maximal dangeruos sets X, Y, we have $g(X \cap Y) = k$ and $g(X \cup Y) = k + 2$,

(3) if X is maximal dangerous and Y is critical then either $X \cap Y = \emptyset$ or $Y \subseteq X$,

(4) for every positive node x there exists a unique maximal critical set S_x with $x \in S_x$,

(5) for two sets S_x, S_y either $S_x = S_y$ or $S_x \cap S_y = \emptyset$ holds,

(6) if $D_1, ..., D_m$ $(m \geq 2)$ are distinct maximal dangerous sets containing a positive node x then $D_i \cap D_j = S_x$ for every $1 \leq i < j \leq m$,

(7) if $D_1, ..., D_m$ $(m \geq 2)$ are distinct maximal dangerous sets containing a positive node x then $g(C) \leq k + m$, where $C := \cup_1^m D_i$. □

Note that $S_x = S_y$ may hold for different positive nodes $x \neq y \in V$. Focus on a fixed pair $x, y \in F$. If there exists a dangerous set X with property (i) above then let M_{xy} be defined as (an arbitrarily fixed) maximal dangerous set with property (i). By Lemma 5(3) we have $S_x, S_y \subseteq M_{xy}$ in this case. If no such set exists then there exist critical sets satisfying property (iia) or (iib). Clearly, in this case $S_x \cap S_y = \emptyset$ and by Lemma 5(1) the union of type (iia) sets (type (iib) sets) with respect to a fixed node pair x, y is also of type (iia) (type (iib), respectively). Thus the maximal type (iia) set, if exists, is equal to S_x and the maximal type (iib) set, if exists, is equal to S_y. Moreover, either $(S_y \cap F) \subseteq \Gamma(S_x)$ or $(S_x \cap F) \subseteq \Gamma(S_y)$.

Lemma 6. *For every $x \in F$ the critical set S_x induces a connected subgraph in G. Suppose $S_x \neq S_y$ for $x, y \in F$ and let M_{xy} be a maximal dangerous set with $x, y \in M_{xy}$. Then either (a) M_{xy} induces a connected subgraph in G or (b) $M_{xy} = S_x \cup S_y$, $S_x \cap S_y = \emptyset$, $\Gamma(M_{xy}) = \Gamma(S_x) = \Gamma(S_y)$, $|\Gamma(M_{xy})| = k - 1$, and $f(S_x) = f(S_y) = 1$.* □

Let us fix an arbitrary positive node x. Since every pair $x, y \in F$ is illegal, using our previous observations we can partition the set F (with respect to x) by defining four sets A, B, C, D as follows:

$A := \{y \in F : S_y = S_x\}$,
$B := \{y \in F : M_{xy}$ exists and $y \notin A\}$,
$C := \{y \in F : M_{xy}$ does not exist and $(S_y \cap F) \subseteq \Gamma(S_x)\}$,
$D := \{y \in F : M_{xy}$ does not exist, $(S_x \cap F) \subseteq \Gamma(S_y)$ and $y \notin C\}$.

Note that for two nodes z, y belonging to different parts of this partition we must have $S_z \cap S_y = \emptyset$. Furthermore, if z belongs to C or D and M_{xy} exists for some y then $S_z \cap M_{xy} = \emptyset$. These facts follow easily from Lemma 5.

Lemma 7. *Let $z \in D$. Then $|\Gamma(S_x) \cap S_z| \geq f(S_x)$.*

Proof. By definition, there exists a positive node $w \in S_z - \Gamma(S_x)$ and hence $Z := S_z - \Gamma(S_x)$ is non-empty. Since no node of Z is adjacent to S_x, we have $|\Gamma(Z) - S_z| \leq g(S_z) - f(S_z) - f(S_x)$. By (1) and $g(S_z) = k$ this gives $k \leq g(Z) = f(Z) + |\Gamma(Z)| = f(Z) + |\Gamma(Z) - S_z| + |\Gamma(Z) \cap S_z| \leq f(S_z) + k - f(S_z) - f(S_x) + |\Gamma(S_x) \cap S_z|$. This inequality proves the lemma. □

Now assume that x is a positive node for which $f(S_x)$ is maximal.

Lemma 8. *If $f(S_x) \geq 2$ then $|F| \leq 2k - 2$.*

Proof. For a node $y \in C$ we have $(S_y \cap F) \subseteq \Gamma(S_x)$ and hence $y \in \Gamma(S_x)$. If $z \in D$ then by Lemma 7 and the choice of x we have $f(S_z) \leq f(S_x) \leq |\Gamma(S_x) \cap S_z|$, hence $f(D) \leq |\Gamma(S_x) \cap (\cup_{z \in D} S_z)|$. Thus $|C| + |D| \leq |Q|$, where $Q := \Gamma(S_x) \cap \cup_{z \in C \cup D} S_z$. Recall that no contribution was counted twice since, as we remarked, two sets S_y, S_z are disjoint whenever y and z belong to different partition classes. Furthermore, $Q \cap M_{xy} = \emptyset$ for each maximal dangerous set M_{xy}. Now let us estimate the contribution of B to $|\Gamma(S_x)|$. If $B = \emptyset$, let $W := S_x$ and $m := 0$, otherwise let W be the union of m distinct maximal dangerous sets, each of the form M_{xy} for some $y \in B$ and such that $B \subset W$. Note that each positive node in $A \cup B$ contributes to $f(W)$. Now $g(W) = f(W) + |\Gamma(W)| \leq k + m$ by Lemma 5 (7). Moreover, by Lemma 6 and $f(S_x) \geq 2$ each maximal dangerous set M_{xy} induces a connected subgraph and hence by Lemma 5 (6) $|\Gamma(S_x) \cap W| \geq m$ holds. This gives $m \leq k - f(S_x) - f(C \cup D)$. Thus $f(W) \leq k + k - f(S_x) - f(C \cup D)$, which yields $|F| = f(A \cup B \cup C \cup D) = f(W) + f(C \cup D) \leq 2k - 2$. □

In the rest of our investigations assume that $f(S_x) = 1$ for every $x \in F$. If every maximal dangerous set M_{xy} (defined with respect to some fixed $x \in F$) induces a connected subgraph, the proof of Lemma 8 works without any modification (except that $f(S_x) \geq 2$ cannot be used in the last count) and gives $|F| \leq 2k - 1$.

Let us fix a node $x \in F$ again and define the partition of F with respect to x as before. Focus on set B, which contains those nodes $y \in F$ for which a maximal dangerous set M_{xy} exists. Let us call such an M_{xy} *special* if it satisfies Lemma 6 (b). Let $B_s := \{y \in B : M_{xy}$ is special$\}$ and let $B_n := B - B_s$. (Note that since $S_x, S_y \subset M_{xy}$ and for a special M_{xy} we have $S_x \cup S_y = M_{xy}$, the set M_{xy} is unique if it is special. Thus the bipartition $B_s \cup B_n$ does not depend on the choices of the M_{xy}'s.) By our previous remark we may assume $B_s \neq \emptyset$.

First suppose that $B_n \neq \emptyset$ and take $z \in B_s, w \in B_n$. We claim that there exists an edge between S_z and $M_{xw} - S_x$. Indeed, since by the definition of B_n and Lemma 6 S_x has a neighbour p in $M_{xw} - S_x$ and since M_{xz} is special, p must be a neighbour of S_z, too. Let W be the union of m distinct maximal non-special dangerous sets M_{xy} such that $B_n \subset W$ and let $R := \cup_{z \in B_s} S_z$. Our claim implies that $|\Gamma(W) \cap R| \geq |B_s|$. Moreover, $R \cap (S_x \cup \Gamma(S_x)) = \emptyset$ holds. Hence, as in Lemma 8, we obtain $f(W) \leq k + m - |\Gamma(W) \cap R| \leq k + k - f(S_x) - f(C \cup D) - |B_s|$. Thus $|F| \leq 2k - 1$ follows.

Now suppose that $B_n = \emptyset$. In this case $|F| = |A| + |C| + |D| + |B_s| \leq f(S_x) + |\Gamma(S_x)| + |B_s| = k + |B_s|$ and therefore $|B_s| \leq k$ yields $|F| \leq 2k$. Suppose $|B_s| \geq k + 1$. By Lemma 6 the special dangerous sets of the form M_{xy} and the maximal critical sets S_x, S_y ($y \in B_s$) have a common set K of neighbours (of size $k - 1$). Moreover, $f(S_y) = 1$ for each $y \in F$. These facts imply that in G' there exist k node-disjoint paths between each pair of nodes x, y of the form $x, y \in K \cup \{r\}$ or $x \in K \cup \{r\}, y \in S_x \cup T$, where $T := \cup_{z \in B_s} S_z$. This proves that the set $K \cup \{r\} \cup S_x \cup T$ induces a k-connected subgraph H in G'. It is easy to see that in this subgraph $K' := K \cup \{r\}$ is a cut of size k, the components of $H - K'$ are precisely the sets S_x and S_z ($z \in B_s$), $d_H(r) = |B_s| + 1 \geq k + 2$ and every edge incident to r is critical with respect to k-connectivity. Finally, we show that this structure implies $G' = H$. To see this suppose that $J := G - (K \cup S_x \cup T) \neq \emptyset$. Since $S_x \cup T$ has no neighbours in J, $|\Gamma(J)| \leq k - 1$ and hence by (1) $f(J) \geq 1$ follows. Let $j \in F \cap J$. By our assumptions $j \in C \cup D$ must hold. Therefore $S_j \cap K \neq \emptyset$ and $S_j \cap (S_x \cup T) = \emptyset$. Since each node in K is adjacent to a node in every S_z ($z \in B_s$), this implies $k \geq |\Gamma(S_j)| \geq |B_s| \geq k + 1$, a contradiction. This proves the theorem via Lemma 4. $\square$

Remark: Consider the complete bipartite graph $K_{k,n-k} = (A, B, E)$ with $|A| = k$ and put $r \in A$. Now $d(r)$ may be arbitrary and there is no admissible pair of edges on r. This shows that there is no lower bound on $d(r)$ which can guarantee the existence of an admissible pair incident to r. To illustrate that in Theorem 4 one has to deal with a more general problem than in Theorem 3, note that if there exists an admissible pair adjacent to r and $d(r) \geq 2k + 1$ in Theorem 3 then any fixed edge rv is part of an admissible pair. This fact was pointed out in [7], where a strongly related "augmentation problem" was considered. However, there are examples showing that such a strengthening of Theorem 4 fails even if $2k + 1$ is replaced by $2k^2 - 2k$.

Finally we remark that Theorem 4 implies Theorem 3 in the case of $d(r) \geq 2k + 1$. This follows by observing that if we split off a pair of edges from a node r in a k-connected graph, preserving k-outconnectivity from r, we preserve k-connectivity, as well.

3 Approximation results for minimum k-outconnected subgraphs

3.1 Uniform cost problems

Our first goal is to give an approximation guarantee less than 2 for the following problem.

Problem P1: Given a graph G, and a node r, find a minimum-size subgraph that is k-outconnected from r.

Our improved approximation guarantees for min-size k-outconnected subgraphs are based on Theorem 1 and results from [3]. We obtain a $(1 + \frac{1}{k})$-approximation algorithm for Problem P1 as follows: First, we find a minimum-size subgraph with minimum degree $(k - 1)$, call it (V, M). This can be done in polynomial time via Edmonds' maximum matching algorithm, or more directly via a b-matching algorithm (i.e., an algorithm for the degree-constrained subgraph problem) [6, Section 11]. Second, we augment (V, M) by an inclusion-wise minimal edge set $F \subseteq E(G) - M$ to obtain a subgraph $H = (V, M \cup F)$ that is k-outconnected from r. Note that every edge $f \in F$ is critical for the k-outconnectivity of H from r. Now, we apply Theorem 1 to H and F and conclude that F is a forest. (Otherwise, if F contains a cycle C, then every node of C is incident to $\geq (k + 1)$ edges of H, since the node is incident to $\geq (k - 1)$ M-edges and to 2 F-edges, but this contradicts Theorem 1.) Therefore, $|F| \leq n - 1$.

Moreover, we have $|M| \leq opt - \lfloor n/2 \rfloor$, by [3, Theorem 3.5]. In detail, [3, Theorem 3.5] shows that for every n-node k-edge connected graph, $\lfloor n/2 \rfloor$ is a lower bound on the difference between the size of the graph and the minimum size of a subgraph of minimum degree $(k-1)$. Clearly, the optimal k-outconnected subgraph of G, call it H^*, is k-edge connected, hence [3, Theorem 3.5] applies to H^* and shows that $|M| \leq$ (minimum size of a subgraph of minimum degree $(k-1)$ of $H^*) \leq opt - \lfloor n/2 \rfloor$. Hence, $|M \cup F| \leq opt + (n/2) \leq (k+1)opt/k$. Here, we used the fact that $opt \geq nk/2$, since the optimal subgraph has minimum degree k.

This method can be extended to the following multi-root version of Problem P1.

Problem P2: Given a graph G and a set $R = \{r_1, \ldots, r_q\}$ of root nodes (more formally, an ordered tuple R of nodes) together with a requirement k_i for each root node r_i. Find a minimum-size subgraph that simultaneously satisfies the connectivity requirements of all the root nodes, that is, the solution subgraph must be k_i-outconnected from r_i, for $i = 1, \ldots, q$.

We use k to denote $\max\{k_i \mid i = 1, \ldots, q\}$, and $\mathbf{k}$ to denote the vector of connectivity requirements $(k_1, \ldots, k_q)$, where q denotes the number of roots, $|R|$. By Lemma 1 there is *no* loss of generality in taking the number of roots q to be $\leq k = \max\{k_i\}$, and moreover, if $q = k$ then $k_1 = \ldots = k_q = k$ and the problem becomes that of finding a minimum-size k-connected spanning subgraph.

We achieve an approximation guarantee of $\min\{2, 1 + \frac{2q-1}{k}\}$ for Problem P2. The approximation guarantee for the multi-root problem is obtained by combining the solution method of Problem P1 and the sparse certificate for "local node connectivity" of Nagamochi & Ibaraki [11]. For a graph H' and nodes v, w, let $\kappa_{H'}(v, w)$ denote the maximum number of openly disjoint $v \leftrightarrow w$ paths. Recall that [11] gave an efficient algorithm for finding j edge disjoint forests $F_1, \ldots, F_j$ of G such that for every two nodes v, w, $\kappa_H(v, w) \geq \min\{j, \kappa_G(v, w)\}$, where the edge set of H is $F_1 \cup \ldots \cup F_j$. This graph H has $\leq k(n-1)$ edges, while the optimal subgraph has $\geq kn/2$ edges. Now H has k_i openly disjoint $v \leftrightarrow r_i$ paths $\forall v \in V, \forall r_i \in R$, because, by assumption, G has k_i openly disjoint $v \leftrightarrow r_i$ paths. Consequently, H is $\mathbf{k}$-outconnected from R, as required, and has size $\leq 2opt$.

Now consider the $\frac{k+2q-1}{k}$-approximation algorithm. First we find a minimum-size subgraph of minimum degree $(k-1)$, call it (V, M). Then, sequentially for each of the roots $r_i = r_1, \ldots, r_q$, we find an inclusionwise minimal edge set $F_i \subseteq E(G) - (F_1 \cup \ldots \cup F_{i-1})$ such that $(V, M \cup F_1 \cup \ldots \cup F_i)$ is k_i-outconnected from r_i. Clearly, the solution subgraph $H = (V, M \cup F_1 \cup \ldots \cup F_q)$ satisfies all the connectivity requirements, i.e., H is $\mathbf{k}$-outconnected from R. By Theorem 1, each F_i ($i = 1, \ldots, q$) has size $\leq (n-1)$, since each F_i is a forest. Also, we have $|M| \leq opt - \lfloor n/2 \rfloor$, by [3, Theorem 3.5]. Hence, $|E(H)| = |M \cup F_1 \cup \ldots \cup F_q| \leq opt + (2q-1)n/2 \leq (k+2q-1)opt/k$, since $opt \geq kn/2$ (since the optimal subgraph has minimum degree k). This proves the following result.

Theorem 5. *There is a polynomial* $\min\{2, \frac{k+2q-1}{k}\}$*-approximation algorithm for the problem of finding a minimum-size subgraph that is k-outconnected from a set of root nodes R, where $q = |R|$ and $k = \max\{k_i\}$.* $\square$

Remark: Note that for $|R| = q \leq k/2$, the approximation guarantee is less than 2, and when k is large and q is small then the approximation guarantee is significantly smaller than 2. In the case of $q = 1$ (Problem P1) the approximation factor equals $1 + \frac{1}{k}$.

3.2 Metric cost problems

Problem P3: Let G be a complete graph, and let $c : E(G) \rightarrow \Re_+$ assign nonnegative costs to the edges such that c is a metric, that is, the edge costs c satisfy the triangle inequality. Given a set of root nodes $R = \{r_1, \ldots, r_q\}$ and connectivity requirements $k = (k_1, \ldots, k_q)$, Problem P3 is to find a min-cost subgraph that is k-outconnected from R.

As we remarked, for the generalization of Problem P3 where the edge costs are nonnegative but arbitrary a $2q$-approximation algorithm is straightforward by the result of Frank & Tardos [4], and by the 2-approximation algorithm of Khuller & Raghavachari [9] for the single-root version of the problem. We will not deal with the general problem but focus on the metric version and improve the approximation factor to 4. Our result is related to [9, Theorem 4.8], but neither result implies the other one. (Theorem 4.8 in [9] gives an approximation guarantee of $(2 + 2(k - 1)/n)$ for the min-cost k-node connected spanning subgraph problem, assuming metric edge costs.) Note that Problem P3 is also NP-hard.

Theorem 6. *Suppose that the edge costs satisfy the triangle inequality. Then there is a polynomial $(3 + \frac{k_s}{k})$-approximation algorithm for the problem of finding a min-cost subgraph that is k-outconnected from the set of root nodes R, where $k := \max\{k_i\}$ is the largest and k_s is the second largest requirement.*

Proof. We start by finding a subgraph H that is k-outconnected from r, with cost $c(H) \leq 2\, opt$. This is done via the Frank-Tardos result, as in [9]. By deleting edges if necessary we may assume that each edge incident to r is critical with respect to k-outconnectivity. If $d_H(r) \geq 2k + 1$ then by Theorem 4 either H is k-connected or we can split off a pair of edges from r preserving k-outconnectivity. In the former case let H be our solution. Clearly, H satisfies all the requirements and has cost $\leq 2opt$. In the latter case we split off the admissible pair. Since c is a metric, this will not increase the cost of our subgraph. Clearly, the admissible pairs, if exist, can be found in polynomial time. Thus we may assume $d_H(r) \leq 2k$. The next step is to add new edges to H in order to make it k_s-connected and to satisfy all the requirements this way. The final solution will be a supergraph of H where each edge added to H has both ends among the neighbours of r. To find the set of new edges to be added take H and split off edges from r, preserving k_s-outconnectivity, until either $d(r) \leq 2k_s$ holds or the graph becomes k_s-connected.

This can be done by Theorem 4. Let H' be the final graph and let E' be the set of new edges obtained by the splittings. Since $d_H(r) \leq 2k$ and $d_{H'}(r) \geq 2k_s - 1$ and $d_H(r) - d_{H'}(r)$ is even, we have $|E'| \leq 2(k - k_s)/2 = k - k_s$. Let C be the set of neighbours of r in H'. Now augment H' by adding an inclusionwise minimal edge set F such that the resulting graph is k_s-connected and each F-edge has both end nodes in C. Since either H' is already k_s-connected or $d_{H'}(r) \leq 2k_s$, by Mader's theorem (Theorem 2) we can see that F is acyclic and so has size $\leq |C| - 1 \leq 2k_s - 1$. Let our solution be $H'' := H + E' + F$, which is k-outconnected from r and k_s-connected and hence satisfies all the requirements by the choice of k and k_s.

We claim that every edge in $E' \cup F$ (in fact, every edge of the complete graph) has cost $\leq opt/k$. To see this, observe that every solution must be k-edge-connected and hence there exist k edge-disjoint paths between any two nodes u, w. Each of these paths has cost at least $c(uw)$ by the triangle inequality. Thus $c(H'') \leq 2opt + (k - k_s + 2k_s)opt/k = (3 + \frac{k_s}{k})opt \leq 4opt.$ □

References

1. D. Bienstock, E. F. Brickell and C. L. Monma, "On the structure of minimum-weight k-connected spanning networks," *SIAM J. Discrete Math.* **3** (1990), 320–329.
2. B. Bollobás, *Extremal Graph Theory*, Academic Press, London, 1978.
3. J.Cheriyan and R.Thurimella, "Approximating minimum-size k-connected spanning subgraphs via matching," manuscript, Sept. 1996. ECCC TR98-025, see http://www.eccc.uni-trier.de/eccc-local/Lists/TR-1998.html. Preliminary version in *Proc. 37th IEEE FOCS* (1996), 292–301.
4. A.Frank and E.Tardos, "An application of submodular flows," *Linear Algebra and its Applications*, **114/115** (1989), 320–348.
5. G.L.Frederickson and J.Ja'Ja', "On the relationship between the biconnectivity augmentation and traveling salesman problems," *Theor. Comp. Sci.* **19** (1982), 189–201.
6. H. N. Gabow and R. E. Tarjan, "Faster scaling algorithms for general graph matching problems," *Journal of the ACM* **38** (1991), 815–853.
7. T. Jordán, "On the optimal vertex-connectivity augmentation," *J. Combinatorial Theory, Series B* **63** (1995), 8–20.
8. S. Khuller, "Approximation algorithms for finding highly connected subgraphs," in *Approximation algorithms for NP-hard problems*, Ed. D. S. Hochbaum, PWS publishing co., Boston, 1996.
9. S. Khuller and B. Raghavachari, "Improved approximation algorithms for uniform connectivity problems," *Journal of Algorithms* **21** (1996), 434–450.
10. W. Mader, "Ecken vom Grad n in minimalen n-fach zusammenhängenden Graphen," *Archive der Mathematik* **23** (1972), 219–224.
11. H.Nagamochi and T.Ibaraki, "A linear-time algorithm for finding a sparse k-connected spanning subgraph of a k-connected graph," *Algorithmica* **7** (1992), 583–596.
12. Z.Nutov, M.Penn and D.Sinreich, "On mobile robots flow in locally uniform networks," *Canadian Journal of Information Systems and Operational Research* **35** (1997), 197–208.
13. R. Ravi and D. P. Williamson, "An approximation algorithm for minimum-cost vertex-connectivity problems." *Algorithmica* (1997) 18: 21-43.

Lower Bounds for On-line Scheduling with Precedence Constraints on Identical Machines

Leah Epstein

Tel-Aviv University *

Abstract. We consider the on-line scheduling problem of jobs with precedence constraints on m parallel identical machines. Each job has a time processing requirement, and may depend on other jobs (has to be processed after them). A job arrives only after its predecessors have been completed. The cost of an algorithm is the time that the last job is completed. We show lower bounds on the competitive ratio of on-line algorithms for this problem in several versions. We prove a lower bound of $2 - 1/m$ on the competitive ratio of any deterministic algorithm (with or without preemption) and a lower bound of $2 - 2/(m+1)$ on the competitive ratio of any randomized algorithm (with or without preemption). The lower bounds for the cases that preemption is allowed require arbitrarily long sequences. If we use only sequences of length $O(m^2)$, we can show a lower bound of $2 - 2/(m+1)$ on the competitive ratio of deterministic algorithms with preemption, and a lower bound of $2 - O(1/m)$ on the competitive ratio of any randomized algorithm with preemption. All the lower bounds hold even for sequences of unit jobs only. The best algorithm that is known for this problem is the well known List Scheduling algorithm of Graham. The algorithm is deterministic and does not use preemption. The competitive ratio of this algorithm is $2 - 1/m$. Our randomized lower bounds are very close to this bound (a difference of $O(1/m)$) and our deterministic lower bounds match this bound.

1 Introduction

We consider the problem of scheduling a sequence of n jobs on m parallel identical machines. There are precedence constraints between the jobs, which can be given by a directed acyclic graph on the jobs. In this graph each directed edge between jobs j_1 and j_2 indicates that j_1 has to be scheduled before j_2. We consider an on-line environment in which a job is known only after all its predecessors in the graph are processed by the on-line algorithm. Each job j has a certain time requirement w_j (which is known when the job arrives). The cost of an algorithm is the makespan, which is the time in which the last job is completed. This model is realistic since often the running times of specific jobs are known in advance,

---

* Dept. of Computer Science, Tel-Aviv University. E-Mail: lea@math.tau.ac.il.

but it is unknown whether after these jobs, there will be need to perform jobs that depend on some previous jobs.

We compare the performance of on-line algorithms and the optimal off-line algorithm that knows the sequence of jobs and the dependencies graph in advance. We use the competitive ratio to measure the performance of the algorithm.

Denote the cost of the on-line algorithm by C_{on} and the cost of the optimal off-line algorithm by C_{opt}. The competitive ratio of an on-line algorithm is r if for any input sequence of jobs: $C_{on} \leq r \cdot C_{opt}$.

We consider a several models. Consider a job j which is released at time t_1 (t_1 is the time that its last predecessor finished running), and has processing time requirement w_j. In the model without preemption, j has to be processed on one machine for w_j time units, starting at some time t, $t \geq t_1$ till time $t + w_j$. In the model that allows preemption, each running job can be preempted, i.e. its processing may be stopped and resumed later on the same, or on different machine. Thus j still has to be processed for a total of w_j time units, but not necessarily continuously, or on one machine (it cannot be processed on more than one machine at the same time). The algorithms may be either deterministic or randomized. For randomized algorithms the competitive ratio is r if for any input sequence of jobs: $E(C_{on}) \leq r \cdot C_{opt}$. It is also possible to consider special sequences as sequences that consist only of unit jobs, and sequences of bounded length.

Related problems and results: The problem of scheduling a set of tasks on parallel machines has been widely studied in many variants. In the basic problem, a sequence of jobs is to be scheduled on a set of identical parallel machines. The jobs may be independent or have precedence constraints between them. They may all arrive at the beginning or have release times, or arrive according to the precedence constraints. The running times of the jobs can be known in advance (at arrival time) or unknown (till they are finished). It is also possible to allow restarts (a job is stopped, and resumed later from the beginning) or to allow preemptions. The goal is to construct a schedule of minimum length. (There are variants with other cost functions too). All those problems, in their off-line version are NP-hard [6].

The first one to introduce the on-line scheduling problem was Graham [8, 9]. He also introduced the algorithm List Scheduling. This algorithm, each time that some machine is idle, schedules a job that is available (if there exists such a job which was not scheduled yet and already arrived). List Scheduling suits all the above mentioned cases, and has the competitive ratio of $2 - 1/m$. In this paper we show that for our problem, the algorithm is optimal in the deterministic case, and that it is almost optimal for randomized algorithms.

Our deterministic lower bounds build on the paper of Shmoys, Wein and Williamson [15]. They consider the problem of scheduling a sequence of independent tasks on-line. The duration of a job in unknown until it is completed, but there are no precedence constraints between the jobs. They show a lower bound of $2 - 1/m$ on the competitive ratio of any deterministic algorithm without preemption. In this paper we adapt this lower bound to a lower bound of

$2 - 1/m$ for our problem. We also build our lower bounds for deterministic algorithms with preemption on some of the ideas of their lower bound. They also show that the same lower bound holds with preemption, and they show a lower bound of $2 - O(1/\sqrt{m})$ on the competitive ratio of randomized algorithms without preemption. The last lower bound can be also adapted to our problem (even with preemption), but we show a much stronger lower bound for randomized algorithms.

Our results: All the results apply even if the sequences may consist of unit jobs only. Also the number of jobs, the structure, and makespan of the optimal off-line assignment is known in advance in all lower bounds.

We prove the following lower bounds for deterministic algorithms:

- A lower bound of $2 - 1/m$ on the competitive ratio of any deterministic on-line algorithm without preemption (even if the length of the sequence is limited to $O(m^2)$).
- A lower bound of $2 - 2/(m+1)$ on the competitive ratio of any deterministic on-line algorithm that allows preemption (even if the length of the sequence is limited to $O(m^2)$).
- A lower bound of $2 - 1/m$ on the competitive ratio of any deterministic on-line algorithm that allows preemption.

We prove the following lower bounds for randomized algorithms

- A lower bound of $2 - 2/(m+1)$ on the competitive ratio of any randomized on-line algorithm without preemption (even if the length of the sequence is limited to $O(m^2)$).
- A lower bound of $2 - O(1/m)$ on the competitive ratio of any randomized on-line algorithm that allows preemption (even if the length of the sequence is limited to $O(m^2)$).
- A lower bound of $2 - 2/(m+1)$ on the competitive ratio of any randomized on-line algorithm that allows preemption.

The similar results for all the models show that for this problem, neither randomization nor preemption can help in reducing the competitive ratio. Any algorithm would have the competitive ratio of $2 - O(1/m)$ which is very close to the competitive ratio of List Scheduling.

More related work: A summary on results for many variants of on-line scheduling problems appears in [14]. Results for the case that jobs arrive over time appear in [1, 5, 15, 17]. Note that [1, 17] show an algorithm with competitive ratio 3/2 even without preemption, for the case that jobs are independent, and the durations are known in advance (unlike our case in which there is a lower bound of $2 - O(1/m)$ even with preemption). Moreover, if preemption is allowed, there exists a 1-competitive algorithm [7, 10, 13, 16]. Results on scheduling with precedence constraints, but for other types of machine sets or jobs can be found in [3, 4, 11, 12]. There are also some off-line results with precedence constraints in [2]. Note that already for a related set of machines (different speeds) the problem of scheduling with precedence constraints is hard (competitive ratio of $\Omega(\sqrt{m})$ [3, 11]).

Structure of the paper: In section 2 we show lower bounds for deterministic algorithms. In section 3 we show lower bounds for randomized algorithms.

2 Deterministic Algorithms

In this section we show a lower bound of $2 - 1/m$ on the competitive ratio of deterministic algorithms with preemption. We give all the lower bounds in this section in detail, since some of the ideas in the randomized lower bounds build on the deterministic case. We begin with a simple lower bound without preemption. All lower bounds use sequences of unit jobs only.

Theorem 1. *The competitive ratio of any deterministic algorithm without preemption is at least $2 - \frac{1}{m}$. This is true even if all jobs are unit jobs.*

Proof. We use the following sequence: (all jobs are unit jobs). First $m(m-1)+1$ jobs arrive. Consider the on-line assignment. Since the total time to schedule $m(m-1)$ jobs is at least $m-1$ units of time, there is at least one job j assigned at time $m-1$ or later, and finishes at time m or later.

After that, a sequence of $m-1$ jobs $j_1, ..., j_{m-1}$ arrives. In this sequence the first job j_1 depends on j, and each job j_i depends on the previous one j_{i-1}; $(2 \leq i \leq m-1)$. It takes more $m-1$ units of time to schedule them and thus $C_{on} = m + (m-1) = 2m - 1$.

The optimal off-line algorithm would schedule j at time 0, and each j_i at time i and thus can finish all m^2 jobs in m time units. Thus $C_{opt} = m$ the competitive ratio is $2 - 1/m$.

We use a longer sequence to show that the same lower bound holds with preemption too.

Theorem 2. *The competitive ratio of any deterministic algorithm with preemption is at least $2 - \frac{1}{m}$. This is true even if all jobs are unit jobs.*

Proof. For an integer k, $(m^k+1)(m-1)+1$ jobs arrive. The minimum time to run those jobs even with preemption is at least $(m^{k+1}+m-m^k)/m = m^k +1-m^{k-1}$, thus for the on-line algorithm there is at least one job j that finishes at time at least $m^k + 1 - m^{k-1}$.

After those jobs, a sequence J of m^k jobs, that the first one depends on j, and each job depends on the previous one arrives. The time to run those jobs is at least m^k and thus $C_{on} = m^k + 1 - m^{k-1} + m^k = 2m^k - m^{k-1} + 1$.

The optimal off-line algorithm would schedule j at time 0, and since there are $m(m^k + 1)$ jobs, the total time to run them would be $C_{opt} = m^k + 1$ (it is possible to run j and all jobs of J in $m^k + 1$ time units). The competitive ratio is $2 - \frac{m^{k-1}+1}{m^k+1} = 2 - \frac{1}{m} - \varepsilon_k$ where $\varepsilon_k \to 0$ when $k \to \infty$. Thus the competitive ratio of any deterministic on-line algorithm that allows preemption is at least $2 - 1/m$.

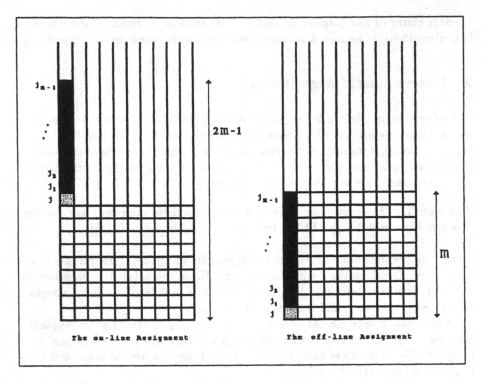

Fig. 1. The on-line and off-line assignments for the sequence in Theorem 1

now we show that the strength of the lower bound almost does not depend on the length of the sequence.

Theorem 3. *The competitive ratio of any deterministic algorithm with preemption is at least* $2 - \frac{2}{m+1}$. *This is true even if all jobs are unit jobs. And the length of the sequence is* $O(m^2)$.

Proof. We use the sequence from Theorem 2 with $k = 1$.

3 Randomized Algorithms

In all the proofs in this section, which are lower bounds on the competitive ratio of randomized algorithms we use an adaptation of Yao's theorem for on-line algorithms. It states that a lower bound for the competitive ratio of deterministic algorithms on any distribution on the input is also a lower bound for randomized algorithms and is given by $E(C_{on}/C_{opt})$. We will use only sequences for which C_{opt} is constant and thus in our case $E(C_{on}/C_{opt}) = E(C_{on})/C_{opt}$.

We begin with a lower bound without preemption. note that in this section the lower bound without preemption uses a totally different sequence than the lower bound with preemption. It is possible the same sequence as in the first proof in this section (Theorem 4) to get the lower bound of $2 - 1/m$ for deterministic algorithms without preemption. Note that here also all lower bound sequences consist of unit jobs only.

Theorem 4. *The competitive ratio of any randomized algorithm without preemption is at least $2 - \frac{2}{m+1}$. This is true even if all jobs are unit jobs.*

Proof. First $m - 1$ phases of $m + 1$ jobs arrive. In each phase, all jobs depend on m jobs of the previous phase (For each phase, the subset of m jobs from the previous phase is chosen among all $\binom{m+1}{m} = m + 1$ possible subsets of m jobs with equal probability). After all $m^2 - 1$ jobs have arrived, another job that depends on m jobs of the last phase arrives (here also the m jobs are chosen among the $m + 1$ possible subsets with equal probability).

For each phase i, $0 \leq i \leq m - 2$, the optimal off-line algorithm schedules the m jobs that the next phase depends on them at time i. all other jobs are scheduled at time $m - 1$ and thus $C_{opt} = m$.

Note that a phase can arrive only after all m jobs from the previous phase, that this phase depends on them were scheduled (the important jobs in this phase). The time to schedule a phase is 1 or 2 units of time, since if the correct subset of m jobs is chosen, it is possible to assign the last job later and use only one time unit. If the wrong subset was chosen, the algorithm has to use another unit of time to complete the running of the important jobs of this phase and uses two units of time. The last job requires exactly one unit of time. Each of the $m - 1$ phases is placed correctly with probability $1/(m+1)$. For each phase, the probability to use a second unit of time is at least $m/(m + 1)$. Thus the expectation of the on-line cost is at least $E(C_{on}) \geq m + (m - 1)(m/(m + 1))$, $C_{opt} = m$ the competitive ratio is at least $2m/(m + 1) = 2 - 2/(m + 1)$.

Now we show the simple lower bound for randomized algorithms with preemption. This lower bound uses a short sequence.

Theorem 5. *The competitive ratio of any randomized algorithm with preemption is at least $2 - O(\frac{1}{m})$. This is true even if all jobs are unit jobs, and the length of the sequence is $O(m^2)$.*

Proof. First m^2 jobs arrive. More m jobs $j_1, ..., j_m$ arrive so that j_1 depends on a subset J of m of the m^2 jobs, which is chosen uniformly at random among all $\binom{m^2}{m}$ possible subsets, and for $1 \leq i \leq m - 1$, j_{i+1} depends on j_i.

Even with preemption, since each job requires one processing unit, and jobs cannot run simultaneously, no jobs can finish before time 1 and at most m jobs can finish at time 1. In general, at most im jobs can finish before time $i + 1$. Let $b_1, ..., b_{m^2}$ be the set of the first m^2 jobs sorted according to their finishing time (b_1 finishes first). For $1 \leq i \leq m$, let J_i be the set $b_{im-m+1}, ..., b_{im}$. Each job in a set J_i cannot finish before time i. Let I be the set of indices i such that there is at least one job of J in J_i. Let i_1 be the maximum index in I. We define p_i to be the probability that $i = i_1$ for each $1 \leq i \leq m$. For $0 \leq i \leq m$, let q_i be the probability that all m jobs of J are chosen among the jobs in the sets $J_1, ..., J_i$ then $q_i = \binom{mi}{m} / \binom{m^2}{m}$ and $p_i = q_i - q_{i-1}$. Let us calculate $E(C_{on})$. For a fixed value of i_1 the on-line cost is at least $i_1 + m$.

$$E(C_{on}) \geq \sum_{i=1}^{m} p_i(i + m) = m \sum_{i=1}^{m} p_i + \sum_{i=1}^{m} i(q_i - q_{i-1})$$

$$= m + \sum_{i=1}^{m} iq_i - \sum_{i=1}^{m} i(q_{i-1}) = m + mq_m + \sum_{i=1}^{m-1} iq_i - \sum_{i=0}^{m-1} (i+1)q_i$$

$$= 2m - (\sum_{i=1}^{m-1} q_i) - q_0 = 2m - \sum_{i=1}^{m-1} q_i$$

(Since $\sum_{i=1}^{m} p_i = 1$, $q_m = 1$ and $q_0 = 0$.) Let us bound the values q_i:

$$q_i = \frac{\binom{mi}{m}}{\binom{m^2}{m}} = \frac{(mi)!(m^2 - m)!}{(m^2)!(mi - m)!} \leq (\frac{i}{m})^m \leq e^{i-m}$$

Thus

$$\sum_{i=0}^{m-1} q_i \leq \frac{e(e^{m-1} - 1)}{e^m(e - 1)} \leq \frac{1}{e - 1}$$

and $E(C_{on}) >= 2m - 1/(e - 1)$. The optimal off-line algorithm would assign all jobs in J at time 0, and get $C_{opt} = m + 1$. Thus the competitive ratio is $2 - O(1/m)$.

We combine the previous lower bound and the deterministic lower bound with preemption to get the following lower bound:

Theorem 6. *The competitive ratio of any randomized algorithm with preemption is at least $2 - \frac{2}{m+1}$. This is true even if all jobs are unit jobs,*

Proof. For an integer k, $N = (m^k + 1)(m - 1) + 1$ jobs arrive. Denote $L = N/m = m^k + 1 - m^{k-1}$. More m^k jobs: $j_1, ..., j_{m^k}$ arrive so that j_1 depends on a subset J of m of the N jobs, which is chosen uniformly at random among all $\binom{N}{m}$ possible subsets, and for $1 \leq i \leq m^k - 1$, j_{i+1} depends on j_i and each job in the new sequence m jobs depends on the previous one.

Even with preemption, since each job requires one processing unit, and jobs cannot run simultaneously, at most im jobs can finish before time $i + 1$. Let $b_1, ..., b_N$ be the set of the first N jobs sorted according to their finishing time (b_1 finishes first). For $1 \leq i \leq L$, let J_i be the set $b_{im-m+1}, ..., b_{im}$. Each job in

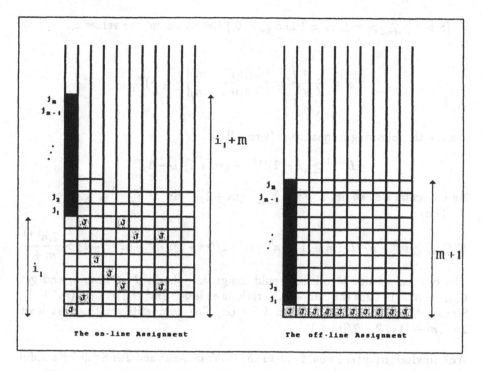

Fig. 2. The on-line and off-line assignments for the sequence in Theorem 5

a set J_i cannot finish before time i. Let I be the set of indices i such that there is at least one job of J in J_i. Let i_1 be the maximum index in I. We define p_i to be the probability that $i = i_1$ for each $1 \leq i \leq L$. For $0 \leq i \leq L$, let q_i be the probability that all m jobs of J are chosen among the jobs in the sets $J_1, ..., J_i$ then $q_i = \binom{mi}{m} / \binom{N}{m}$ and $p_i = q_i - q_{i-1}$. Let us calculate $E(C_{on})$. For a fixed value of i_1 the on-line cost is at least $i_1 + m^k$.

$$
E(C_{on}) \geq \sum_{i=1}^{L} p_i(i + m^k) = m^k \sum_{i=1}^{L} p_i + \sum_{i=1}^{L} i(q_i - q_{i-1})
$$

$$
= m^k + \sum_{i=1}^{L} iq_i - \sum_{i=1}^{L} i(q_{i-1}) = m^k + Lq_L + \sum_{i=1}^{L-1} iq_i - \sum_{i=0}^{L-1} (i+1)q_i
$$

$$
= m^k + L - (\sum_{i=1}^{L-1} q_i) - q_0 = m^k + L - \sum_{i=1}^{L-1} q_i
$$

(Since $\sum_{i=1}^{L} p_i = 1$, $q_L = 1$ and $q_0 = 0$.) Let us bound the values q_i:

$$q_i = \frac{\binom{mi}{m}}{\binom{N}{m}} = \frac{(mi)!(N-m)!}{(N)!(mi-m)!} \leq \left(\frac{i}{L}\right)^m$$

We use the following inequality of Bernoulli:

$$L^{m+1} \geq (L-1)^{m+1} + (m+1)(L-1)^m$$

By induction we can get $L^{m+1} \geq \sum_{i=1}^{L-1}(m+1)i^m$ hence $\sum_{i=1}^{L-1} q_i \leq \frac{L}{m+1}$.
Thus

$$E(C_{on}) \geq m^k + L - L/(m+1) = m^k + (1 - 1/(m+1))(m^k - m^{k-1} + 1) > \frac{2m^{k+1}}{m+1}$$

The optimal off-line algorithm would assign all jobs in J at time 0, and get $C_{opt} = m^k + 1$. Thus the competitive ratio is at least $(2m^{k+1})/((m+1)(m^k+1))$. Since $m^k/(m^{k+1} + 1) \to 1$ when $k \to \infty$, The competitive ratio is at least $2m/(m+1) = 2 - 2/(m+1)$

Acknowledgments: I would like to thank Yossi Azar and Jiří Sgall for helpful discussions.

References

1. B. Chen and A. Vestjens. Scheduling on identical machines: How good is lpt in an on-line setting? *To appear in Oper. Res. Lett.*
2. Fabian A. Chudak, David B. Shmoys. Approximation algorithms for precedence constrained scheduling problems on parallel machines that run at different speeds. *Proc. of the 8th Ann. ACM-SIAM Symp. on Discrete Algorithms*, 581-590, 1997.
3. E. Davis and J. M. Jaffe. Algorithms for scheduling tasks on unrelated processors. *J. ACM.* 28(4):721-736, 1981.
4. A. Feldmann, M.-Y. Kao, J. Sgall and S.-H. Teng. Optimal online scheduling of parallel jobs with dependencies. *Proc. of the 25th Ann. ACM symp. on Theory of Computing*, pages 642-651.
5. A. Feldmann, B. Maggs, J. Sgall, D. D. Sleator and A. Tomkins. Competitive analysis of call admission algorithms that allow delay. Technical Report CMU-CS-95-102, Carnegie-Mellon University, Pittsburgh, PA, U.S.A., 1995.
6. M. R. Garey and D. S. Johnson. *Computers and Intractability: A Guide to the Theory of NP-completeness*, W. H. Freeman, New-York, 1979.
7. T. Gonzalez and D. B. Johnson. A new algorithm for preemptive scheduling of trees. *J. Assoc. Comput. Mach.*, 27:287-312, 1980.
8. R.L. Graham. Bounds for certain multiprocessor anomalies. *Bell System Technical Journal*, 45:1563-1581, 1966.
9. R.L. Graham. Bounds on multiprocessing timing anomalies. *SIAM J. Appl. Math*, 17:263-269, 1969.

10. K. S. Hong and J. Y.-T. Leung. On-line scheduling of real-time tasks. *IEEE Transactions on Computers*, 41(10):1326-1331, 1992.

11. Jeffrey M. Jaffe Efficient scheduling of tasks without full use of processor resources. *The. Computer Science*, 12:1-17, 1980.

12. J.W.S. Liu and C.L. Liu. Bounds on scheduling algorithms for heterogeneous computing systems. *Information Processing 74, North Holland*, 349-353,1974.

13. S. Sahni and Y. Cho. Nearly on line scheduling of a uniform processor system with release times. *Siam J. Comput.* 8(2):275-285, 1979.

14. J. Sgall. On-Line Scheduling - A Survey 1997

15. D. B. Shmoys, J. Wein and D. P. Williamson. Scheduling parallel machines on line. *Siam J. of Computing*, 24:1313-1331, 1995.

16. A. P. A. Vestjens. Scheduling uniform machines on-line requires nondecreasing speed ratios. Technical Report Memorandum COSOR 94-35, Eindhoven University of Technology, 1994. To appear in *Math. Programming*.

17. A. P. A. Vestjens. On-line Machine Scheduling. Ph.D. thesis, Eindhoven University of Technology, The Netherlands, 1997.

Instant Recognition of Half Integrality
and 2-Approximations

Dorit S. Hochbaum

Department of Industrial Engineering and Operations Research, and Walter A. Haas
School of Business, University of California, Berkeley

Abstract. We define a class of integer programs with constraints that
involve up to three variables each. A generic constraint in such integer
program is of the form $ax + by \leq z + c$, where the variable z appears
only in that constraint. For such binary integer programs it is possible
to derive half integral superoptimal solutions in polynomial time. The
scheme is also applicable with few modifications to nonbinary integer
problems. For some of these problems it is possible to round the half
integral solution to a 2-approximate solution. This extends the class of
integer programs with at most two variables per constraint that were
analyzed in [HMNT93]. The approximation algorithms here provide an
improvement in running time and range of applicability compared to
existing 2-approximations. Furthermore, we conclude that problems in
the framework are MAX SNP-hard and at least as hard to approximate
as vertex cover.
Problems that are amenable to the analysis provided here are easily
recognized. The analysis itself is entirely technical and involves manipulating the constraints and transforming them to a totally unimodular
system while losing no more than a factor of 2 in the integrality.

1 Introduction

We demonstrate here for a given class of integer programming problems a unified
technique for deriving constant approximations and superoptimal solutions in
half integers. The class is characterized by formulations involving up to three
variables per constraint one of which appears in a single constraint. Such problem
formulations have a number of interesting properties that permit the derivation
of lower or upper bounds that are of better quality than the respective linear
programming relaxations, and that can be turned in some cases into feasible
solutions that are 2-approximate.

The integer programs described here have a particularly structured set of
constraints, with a generic constraint of the form $ax + by \leq z + c$. Note that
any linear optimization problem, integer or continuous, can be written with at
most three variables per inequality. The generic constraints here have one of

* Research supported in part by NSF award No. DMI-9713482, and by SUN
Microsystems.

the variables appearing only once in a constraint. We call such structure of the set of constraints the 2var structure. The paper describes an analysis of such problems based on manipulating the constraints and transforming them to a totally unimodular system. The outcome of this process is a polynomial algorithm that delivers a superoptimal solution to all integer programs with 2var constraints that has the variables on the left hand side as integer multiple of $\frac{1}{2}$, and the variables on the right hand side, z as an integer multiple of $\frac{1}{2}\frac{1}{U}$ where U is the number of integer values in the range of the variables x and y. In the case of binary problems there is an exception with the value $U = 2$ with z being an integer multiple of $\frac{1}{2}$ (although there are two integer values for each variable). Being superoptimal means that the solution's objective value can only be better (lower) than the optimum (for minimization problems).

When it is possible to round the $\frac{1}{2}$ integral solution to a feasible solution then a 2-approximate solution is obtained.

Applications and 2-approximations based on the technique. The class of problems addressed by the technique described expands substantially the class treated in [HMNT93]. There we demonstrated that $\frac{1}{2}$ integral solutions and 2-approximations are *always* obtained in polynomial time for minimization (with nonnegative coefficients) integer programs with at most *two* variables per inequality. For these problems a feasible rounding always exists provided that the problems are feasible. The technique of [HMNT93] was shown applicable to the vertex cover problem and the problem of satisfying a 2SAT formula with the least weight of true variables. Here and in other papers we demonstrate 2-approximations for additional problems:

- **minimum satisfiability**. In the problem of *minimum satisfiability* or MIN-SAT, we are given a CNF satisfiability formula. The aim is to find an assignment satisfying the smallest number of clauses, or the smallest weight collection of clauses. The minimum satisfiability problem was introduced by Kohli et. al. [KKM94] and was further studied by Marathe and Ravi [MR96]. Marathe and Ravi discovered a 2-approximation algorithm to the problem, that can be viewed as a special case of our general algorithm for problems with two variables per inequality.
- **Scheduling with precedence constraints**, [CH97].
- **Biclique problems**. Minimum weight node deletion to obtain a complete bipartite subgraph – biclique [Hoc97]. With the use of the technique described here we identified the polynomiality of the problem on general graphs when each side of the biclique is not required to form an independent set. When this requirement is introduced the problem is NP-hard and a 2-approximation algorithm is given. Another variant of the biclique problem is the minimum weight edge deletion. This problem is NP-hard even on a bipartite graphs. All variants of this problem considered in [Hoc97] are NP-hard and 2-approximable.
- **The complement of the edge maximum clique problem**. This problem is to minimize the weight of the edges deleted so the remaining subgraph is a clique, [Hoc97].

- **Generalized satisfiability problems**, [HP97].
- **The generalized vertex cover problem**, [Hoc97a].
- **The t-vertex cover problem**. [Hoc98].
- **The feasible cut problem**. The 2-approximation for this problem is illustrated here.

The last three items on the list are problems that have three variables per inequality. The technique delivers $\frac{1}{2}$ integral solutions as well as polynomial time 2- approximation algorithms for these problems. All the approximations listed are obtained using a minimum cut algorithm, and thus in strongly polynomial time.

The problems that fall in the class described here and are NP-hard are then provably at least as hard to approximate as the vertex cover problem (see [HMNT93] and [Hoc96]). Therefore an approximation factor better than 2 is only possible provided that there is such approximation for the vertex cover.

1.1   The Structure of 2var-Constraints

2var constraints contain two types of integer variables, a vector $\mathbf{x} \in \mathbb{N}^n$ and $\mathbf{z} \in \mathbb{N}^{m_2}$. We refer to them here as x-variables and z-variables respectively. A 2var constraint has at most two x-variables appearing in it. The z-variables can appear each in just one constraint.

We refer to integer problems of optimizing over 2var constraints as IP2. A formulation of a typical IP2 is,

$$
\begin{array}{llll}
& \text{Min} & \sum_{j=1}^n w_j x_j + \sum e_i z_i & \\
& \text{subject to} & a_i x_{j_i} + b_i x_{k_i} \geq c_i + d_i z_i & \text{for } i = 1, \ldots, m \\
\text{(IP2)} & & \ell_j \leq x_j \leq u_j & j = 1, \ldots, n \\
& & z_i \text{ integer} & i = 1, \ldots, m_2 \\
& & x_j \text{ integer} & j = 1, \ldots, n.
\end{array}
$$

A parameter of importance in the analysis is the number of values in the range of the x-variables, $U = max_{j=1,\ldots n}(u_j - \ell_j + 1)$. We assume throughout that U is polynomially bounded thus permitting a reference to running time that depends polynomially on U as polynomial running time.

An important property of 2var inequalities that affect the complexity of the IP2 is monotonicity:

Definition 1. *An inequality $ax - by \leq c + dz$ is* **monotone** *if $a, b \geq 0$ and $d = 1$.*

The IP2 problem was studied for $D = \max |d_i| = 0$ in [HN94] and in [HMNT93]. For such problems there are only x-variables and at most two of them per inequality. Hochbaum and Naor [HN94] devised a polynomial time algorithm to solve the problem in integers over *monotone* inequalities. For (nonmonotone) inequalities with at most two variables per inequality, $D = 0$, Hochbaum,

Megiddo, Naor and Tamir described a polynomial time 2-approximation algorithm. These results are extended here for $D \geq 1$.

An important special case of IP2 where the value of U affects neither the complexity nor the approximability is of binarized IP2:

Definition 2. *An IP2 problem is said to be **binarized** if all coefficients of the variables are in $\{-1, 0, 1\}$. Or, if $B = \max_i\{|a_i|, |b_i|, |d_i||\} = 1$.*

Note that a binarized system is not necessarily defined on binary variables. The value of U may be arbitrarily large for a binarized IP2 without affecting the polynomiality of the results as proved in our main theorem.

1.2   The Main Theorem

For IP2 over *monotone* constraints that is binarized we describe a polynomial time algorithm solving the problem in integers. The polynomial algorithm for the monotone IP2 is used as a building block in the derivation of the superoptimal solution and approximations of nonmonotone problems.

For nonmonotone constraints a polynomial time procedure derives a superoptimal half integral solution for the x-variables in which the z-variables are an integer multiple of $\frac{1}{2DU}$. If that solution has a rounding to a feasible integer solution, and the objective function coefficients are nonnegative then that solution is a 2-approximation, or for arbitrary D and U it is a $2DU$-approximation.

For an integer program over 2var constraints, the running time required for finding a superoptimal half integral solution can be expressed in terms of the time required to solve a linear programming over a totally unimodular constraint matrix, or in terms of minimum cut complexity. In the complexity expressions we let $T(n, m)$ be the time required to solve a minimum cut problem on a graph with m arcs and n nodes. $T(n, m)$ may be assumed equal to $O(mn \log(n^2/m))$, [GT88]. For binarized system the running time depends on the complexity of solving a minimum cost network flow algorithm $T_1(n, m)$. We set $T_1(n, m) = O(m \log n(m + n \log n))$, the complexity of Orlin's algorithm [Orl93].

Let $T = T(2 \sum_{j=1}^{n} [u_j - \ell_j], 2m_1(U - 1) + 2m_2 U^2)$ and $T_1 = T_1(2 \sum_{j=1}^{n} [u_j - \ell_j], 2m_1 U + 2m_2(U + 1)^2)$. The main theorem summarizing our results is,

Theorem 3. *Given an IP2 on $m = m_1 + m_2$ constraints, $\mathbf{x} \in \mathbf{Z^n}$ and $U = max_{j=1,\ldots n}(u_j - \ell_j + 1)$.*

1. *A monotone IP2 with $D \leq 1$ is solvable optimally in integers for*
   - *$U = 2$, in time $T(n, m)$.*
   - *Binarized IP2 in time $T_1(n, m)$.*
   - *Otherwise, in time T.*
2. *For nonmonotone IP2 with $D \leq 1$, a superoptimal fractional solution is obtainable in polynomial time:*
   - *For $U = 2$ a half integral superoptimal solution is obtained in time $T(2n, 2m)$.*

- *For binarized IP2, a half integral superoptimal solution is obtained in time $T_1(2n, 2m)$.*
- *Otherwise, a superoptimal solution that is integer multiple of $\frac{1}{2}$ for the x-variables, and integer multiple of $\frac{1}{2U}$ for the z-variables is obtained in time T.*

3. *Given an IP2 and an objective function $\min \mathbf{wx} + \mathbf{cz}$ with $\mathbf{w}, \mathbf{c} \geq \mathbf{0}$.*
   - *For $D = 0$, if there exists a feasible solution then there exists a feasible rounding of the $\frac{1}{2}$ integer solution, which is a 2-approximate solution obtainable in time T, [HMNT93].*
   - *For binarized IP2, if there exists a feasible rounding of the fractional solution, then any feasible rounding is a 2-approximate solution obtainable in time T_1.*
   - *For $D > 0$, if there exists a feasible rounding of the fractional solution, then any feasible rounding is a $2DU$-approximate solution obtainable in time T. If $U = 2$ the solution is $2D$-approximate.*

2 The Algorithm

The algorithm solves the monotone IP2 problem in polynomial time using minimum cut or maximum flow algorithm. This is done by transforming the monotone constraints to an equivalent system that is totally modular but contain a larger number of constraints by a factor of U.

For NP-hard instances of IP2, those containing nonmonotone constraints, we employ a *monotonizing* procedure. This procedure is a transformation of the 2var constraints to another set of constraints with totally unimodular matrix coefficients. The transformed problem can then be solved in integers. The transformation back to the original polytope maps integers to half integers for the x-variables. Those can be rounded, under certain conditions, to a feasible solution within a factor of 2 of the optimum.

Algorithm IP2 described in Figure 1 works in two phases: Phase applies a process of "monotonizing" (step 1) the inequalities and "binarizing" (step 2) the variables. The second phase recovers the values of the fractional solution that is half integral for the x-variables and an integer multiple of $\frac{1}{2DU}$ for the z-variables.

2.1 Monotonizing

Consider first a generic nonmonotone inequality $ax + by \leq c + dz$. It can be assumed that z is scaled so that $d > 0$ and its objective function coefficient is positive. If the inequality is reversed, $ax + by \geq c + dz$, z is simply set to its lower bound. Replace each variable x by two variables, x^+ and x^-, and each term dz by z' and z''. The nonmonotone inequality is then replaced by two monotone inequalities:

$$ax^+ - by^- \leq c + z', \quad -ax^- + by^+ \leq c + z''.$$

The upper and lower bounds constraints $\ell_j \leq x_j \leq u_j$ are transformed to

---

Algorithm IP2 (min{ay : By ≤ c})

1. **Monotonize** using the map $f : \mathbf{y} \to \mathbf{y}^+, \mathbf{y}^-$, $B'\mathbf{y} \le \mathbf{c}'$.
2. **Binarize**, $B''\mathbf{y} \le 0$
3. **Solve** min{a'y : B''y ≤ 0} in integers. Let optimal solution be $\hat{\mathbf{y}}^+, \hat{\mathbf{y}}^-$.
4. **Recover** fractional solution $\hat{\mathbf{y}}$ to {By ≤ c} by applying $f^{-1}(\hat{\mathbf{y}}^+, \hat{\mathbf{y}}^-)$.
5. **Round.** If a feasible rounding exists, round $\hat{\mathbf{y}}$ to $\mathbf{y}^*$.

Fig. 1. Schematic description of the algorithm IP2

$$\ell_j \le x_j^+ \le u_j \qquad -u_j \le x_j^- \le -\ell_j .$$

In the objective function, the variable x is substituted by $\frac{1}{2}(x^+ - x^-)$ and z is substituted by $\frac{1}{2}(z' + z'')$.

Monotone inequalities remain so by replacing the variables x and y in one inequality by x^+ and y^+, and in the second, by x^- and y^-, respectively. The variable z is duplicated:

$$ax^+ - by^+ \le c + z'$$
$$ax^- - by^- \le c + z'' .$$

It is easy to see that if $x^+, x^-, y^+, u^- z', z''$ solve the transformed system, then $x = \frac{1}{2}(x^+ - x^-), y = \frac{1}{2}(y^+ - y^-), z = \frac{1}{2d}(z' + z'')$ solve the original system.

2.2 Binarizing

The transformed monotonized and binarized system has the property that every extreme point solution is integer. The process of conversion of the system of inequalities to a system that has all coefficients in $\{0, -1, 1\}$ is referred to as *binarizing*. This process can be applied to a 2var system of inequalities whether or not it is monotonized. For simplicity we describe the binarizing of monotonized inequalities.

"Binarizing" is a process transforming the monotonized system to inequalities of the form

$$x_i - x_j \le 0 \text{ or } x_i - x_j \le z_{ij} .$$

We start by replacing each x-variable $x_i, \ell \le x_i \le u$, by $u - \ell$ binary variables $x_i^{(\ell+1)}, \ldots, x_i^{(u)}$ so that $x_i = \ell + x_i^{(\ell+1)} + x_i^{(\ell+2)} + \ldots + x_i^{(u)}$, and so that $x_i \ge k x_i^{(k)}$. The values of $x_i^{(\ell+1)}, \ldots, x_i^{(u)}$ form a consecutive sequence of 1s followed by a consecutive sequence of 0s. Each of these sequences could possibly be empty. This structure is enforced by including for each variable x_i the inequalities,

$$x_i^{(k)} \ge x_i^{(k+1)} \qquad \text{for } k = \ell, \ldots, u - 1, \tag{1}$$

where $x_i^{(\ell)} = 1$.

For a linear objective function with x_i's coefficient w, the term wx_i is replaced by,

$$w\ell + wx_i^{(\ell+1)} + \ldots + wx_i^{(u)}.$$

Binarizing $ax_i - bx_j \leq c$: Consider the monotone inequality $ax_i - bx_j \leq c$. This inequality enforces for each value $p_i \in [\ell_i, u_i]$ the implication: if $x_i \geq p_i$, x_j must satisfy for p_j a function of p_i,

$$x_j \geq \lceil \frac{ap_i - c}{b} \rceil \equiv p_j. \tag{2}$$

In other words, the implications

$$x_i^{(p_i)} = 1 \implies x_j^{(p_j)} = 1 \text{ for all } p_i \in [\ell_i, u_i]$$

are equivalent to the inequality $ax_i - bx_j \leq c$. If $p_j > u_j$ then the upper bound on x_i is updated by setting, $x_i \leq p_i - 1$ and $u_i \leftarrow p_i - 1$. To satisfy the set of implications it suffices to include the inequalities, $x_j^{(p_j)} \geq x_i^{(p_i)}$. We append the set of inequalities (1) with up to $\min\{u_i, u_j\}$ inequalities, $\forall p_i \in \{\ell_i, \ldots, u_i\}$ such that $p_j \in \{\ell_j, \ldots, u_j\}$,

$$x_j^{(p_j)} \geq x_i^{(p_i)}. \tag{3}$$

The set of inequalities (3) is equivalent to the inequality $ax_i - bx_j \geq c$.
Binarizing $ax_i - bx_j \leq dz + c$: Consider an inequality with three variables, $ax_i - bx_j \leq dz' + c$. First we substitute dz' by another variable z, thus deriving the inequality $ax_i - bx_j \leq z + c$. We assume that z's coefficient in the objective function is positive, else we can fix z at its upper bound, u_z, in an optimal solution.

Since the procedure is somewhat involved, we describe it here only for the case when z is binary. The general case is described in [Hoc97a].
z binary: Let p_j be a function of p_i as in (2). If $x_i^{(p_i)} = 1$ and

$$p_j - 1 = \lceil \frac{ap_i - c - 1}{b} \rceil < \lceil \frac{ap_i - c}{b} \rceil = p_j,$$

then, if $z = 1$, $x_j^{(p_j)}$ can be 0 (that is $x_j \leq p_j - 1$ rather than $x_j \geq p_j$). If the strict inequality is not satisfied then $z = 0$ in any optimal solution. The condition is enforced by setting $z = 1$ whenever $x_i^{(p_i)} = 1$ and $x_j^{(p_j)} = 0$, or equivalently, by satisfying the inequality $x_i^{(p_i)} - x_j^{(p_j)} \leq z$. Since in the case considered z cannot be great than 1, x_j must be at least $p_j - 1$ and we add the inequality $x_i^{(p_i)} \leq x_j^{(p_j - 1)}$.

Now inequality $x_i^{(p_i)} - x_j^{(p_j)} \leq z$ with z in the right hand side may appear a number of times. There could be as many as U inequalities with z in the right hand side. Such set of inequalities is not totally unimodular. To convert it into a totally unimodular system we treat each occurrence of z as a separate variable,

$z_{(p_i)}$ and set $z = \sum_{\ell_i \leq p_i \leq u_i} z_{(p_i)}$. If there are several inequalities for z with the same p_j value then all but the one with the lowest p_i value are redundant. This argument leads to the conclusion that the number of inequalities can be no larger than U. Consequently, each occurrence of $z_{(p_i)}$ is assigned $\frac{1}{U}$ of the cost of z in the objective function. This guarantees that the objective function is a lower bound on the integer optimum.

2.3 The Network

Here we show that the totally unimodular system of inequality constraints is solvable by a minimum cut algorithm.

Theorem 4. *The integer problem on a monotone binarized system of inequalities is solvable by minimum s,t-cut algorithm on an appropriately defined network.*

The monotonized problem can be solved in integers using linear programming. Instead, we create a network where the source set of a minimum cut corresponds to an optimal solution to the monotonized binarized system.

Each binary variable $x_i^{(k)}$ has an associated node in the graph. For each variable x_i there is a node for the lower bound ℓ_i, $x^{(\ell_i)}$. All lower bound nodes are connected to a dummy source node, s, with infinite capacity arcs. This guarantees that the lower bound nodes are always in the source set of any finite cut. In fact all these nodes can be shrunk with the source and not be present explicitly in the graph. We refer to their presence only to facilitate the presentation.

There is a chain of infinite capacity arcs, each going from $x_i^{(k+1)}$ to $x_i^{(k)}$. We refer to these collections of arcs for each x_i as the x_i-chain. These represent the set of inequalities (1) in the sense that any finite cut that contains any node in the source set also contains all of its successors and thus represent a solution satisfying these inequalities.

Each node has weight which is the objective coefficient associated with its corresponding variable. These weights are permitted to be either positive or negative. The lower bound nodes assume the value $w_i \ell_i$, and every other node in the chain has the weight w_i. There are arcs of capacity $|w|$ from s to each node of negative weight w, and arcs of capacity w from each node of positive weight w to t. With those additional arcs, a minimum s,t-cut is shown to correspond to an optimal solution to the monotonized system as described next.

The set of inequalities (3) is represented by arcs of capacity ∞ from node p_i of x_i to node p_j of x_j. These are shown in Figure 2(a).

The network for an inequality involving z is depicted in Figure 2(b). This includes the case of these variables being non binary. For z binary there is only one arc corresponding to z originating at each node in the chain.

The next Lemma is proved in the full version of this paper.

Lemma 5. *The source set of any finite cut in the network corresponds to a feasible integer solution to the system of inequalities.*

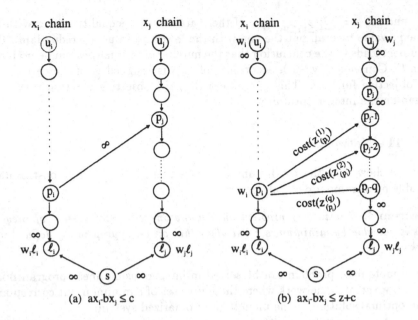

Fig. 2. The network for x_i and x_j

To recover the values of the x and z-variables we set for each node in the source set the corresponding value of 1 to the associated variable, and 0 to those in the sink set. The following equations are then used to recover the values of the x and z variables.

$$x_i = \ell_i + x_i^{(\ell_i+1)} + x_i^{(\ell_i+2)} + \ldots + x_i^{(u_i)},$$

$$z^{(q)} = \sum_{\ell_i \leq p_i \leq u_i} z_{(p_i)}^{(q)}$$

$$z = \ell_z + \frac{1}{U}[(z(1) - \ell_z)z^{(1)} + (z(2) - z(1))z^{(2)} + \ldots + (z(\bar{q}) - z(\bar{q}-1))z^{(\bar{q})}].$$

Lemma 6. *The minimum s,t-cut in the network constructed corresponds to an optimal solution to the monotonized IP2 with the objective* $\min \sum w_j x_j + \sum c_{ij} z_{ij}$.

Proof. From the previous lemma the source set of a finite cut represents a feasible solution. Consider a finite cut (S,T) in the network, and let the sum of all the negative node weights be W, $W = \sum_{w_j < 0}(-w_j)$. Note that W is a constant. Let $C(S,T)$ be the capacity of the cut.

$$
\begin{aligned}
C(S,T) &= \textstyle\sum_{w_j > 0, j \in S} w_j + \sum_{w_i < 0, i \in T}(-w_i) + \sum_{i \in S, j \in T} c_{ij} \\
&= \textstyle\sum_{w_i < 0}(-w_i) - \sum_{w_i < 0, i \in S}(-w_i) + \sum_{w_j > 0, j \in S} w_j + \sum_{i \in S, j \in T} c_{ij} \\
&= W + \textstyle\sum_{j \in S} w_j + \sum_{i \in S, j \in T} c_{ij}.
\end{aligned}
$$

Hence minimizing the cut capacity is equivalent to minimizing the sum of weights of nodes in S plus the capacity of the arcs in the cut separating S from T, which is precisely the value of our objective function. □

3   Minimum Capacity Feasible Cut

The *feasible cut* problem was introduced by Yu and Cheriyan [YC95]. In this problem we are given an undirected graph $G = (V, E)$ with nonnegative edge weights, c_{ij}, k pairs of "commodities" vertices $\{s_1, t_1; \ldots; s_k, t_k\}$ and a specified node $v*$. The problem is to find a partition of V, $(S, \bar{S})$, so that $v* \in S$ and so that for every commodity pair s_ℓ, t_ℓ, has at most one node in S, and such that the cost of the cut $C(S, \bar{S})$ is minimum. Yu and Cheriyan proved that the problem is NP-hard and gave a 2-approximation algorithm. Their algorithm requires to solve a linear program that, as we show here, has optimal solution consisting of half integrals.

Our treatment of the feasible cut problem is applicable to a generalized form of the problem permitting a directed graph and weighted nodes and k commodity sets of nodes $T_1, \ldots, T_k$ where $|T_\ell| \geq 2$ for $\ell = 1, \ldots, k$. The generalized feasible cut problem is to find a partition of the V in a (directed) edge weighted graph $G = (V, A)$, $(S, \bar{S})$, so that $v* \in S$ and so that every commodity set, T_ℓ, has at most one node in S, and the cost of the cut $C(S, \bar{S})$ with the cost of the nodes in the source set is minimum. The undirected version can be formulated as a directed one by replacing each edge $\{i, j\}$ by a pair of arcs (i, j) and (j, i) of equal cost $c_{ij} = c_{ji}$. Let $x_i = 1$ if $i \in S$ and 0 otherwise.

$$
\begin{array}{ll}
\text{Min} & \sum_{(i,j) \in A} c_{ij} z_{ij} + \sum_{j \in V} w_j x_j \\
\text{subject to} & x_i - x_j \leq z_{ij} \quad (i, j) \in A \\
\text{(Feas-Cut)} & x_{p_\ell} + x_{q_\ell} \leq 1 \quad p_\ell, q_\ell \in T_\ell \text{ for } \ell = 1, \ldots, k \\
& x_{v*} = 1 \\
& x_i, z_j \text{ binary for all } i, j.
\end{array}
$$

The constraints in Feas-Cut have the 2var structure. To see that the formulation is valid, observe that for an optimal solution $\mathbf{x}^*$ the set $S = \{j : x_j^* = 1\}$ forms the desired cut. Note that $S \neq V$ as required since all vertices in a commodity set but one, must assume the x value 0. We can now monotonize and solve with linear programming. Better still, we construct a network in which a minimum cut where $v*$ is at the source set of an optimal solution. Figure 3 illustrates the basic gadget in the network for feasible cut. In the figure, each inequality in the first set of inequalities corresponds to the two horizontal arcs, and each inequality in the second set corresponds to the diagonal arcs.

After deriving the $\frac{1}{2}$ integral optimal solution $\hat{\mathbf{x}}, \hat{\mathbf{z}}$ we round it to a feasible solution: The values of $\hat{\mathbf{z}}$ that are $\frac{1}{2}$ are all rounded up to 1 and the values of $\hat{\mathbf{x}}$ that are $\frac{1}{2}$ are rounded down to 0. Denote this feasible rounded solution by $\mathbf{x}^*, \mathbf{z}^*$ and the optimal solution value by OPT. This rounded solution is 2-approximate:

$$
\sum c_{ij} z_{ij}^* + \sum w_j x_j^* \leq 2 \cdot \sum c_{ij} \hat{z}_{ij} + \sum w_j \hat{x}_j \leq 2 \cdot \text{OPT}.
$$

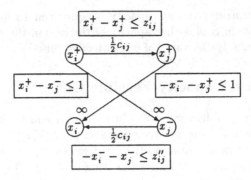

Fig. 3. Feasible cut

As for the complexity, the problem is finding a minimum cut with a specified source on a network with $O(n)$ nodes and $O(M)$ arcs for $M = |A| + \sum^k \binom{|T_\ell|}{2}$. This problem can be solved in the same complexity as a single maximum flow problem in $O(Mn \log \frac{n^2}{M})$ using the algorithm of Hao and Orlin [HO94].

4 Conclusions

Algebraic manipulation of the constraint matrix of some integer programming formulations is shown an effective approximation algorithmic tool. It is hence useful to focus on alternative formulations of problems and perhaps other types of reductions to totally unimodular matrices.

It is intriguing that there are additional half integrality results of [GVY93a] and [GVY94] for multiway directed cuts on edges or nodes (all pairs multicut) and for multicut on trees that we cannot explain with the framework proposed. These might perhaps be explained by reductions to other types of totally unimodular matrices. Another possibility is that there exists another 2var formulation that so far we failed to identify.

There are other problems for which we know of 2-approximations but not of $\frac{1}{2}$ integrality. These include the vertex feedback problem (undirected), the k-center problem and the directed arc feedback set with an objective to maximize the weight of the remaining arcs in the acyclic graph (see [Hoc96a] for descriptions of these problems and approximation algorithms). Are $\frac{1}{2}$ integrality results possible for these problems? At this point the question remains open.

References

[CH97] F. Chudak and D. S. Hochbaum. A half-integral linear programming relaxation for scheduling precedence-constrained jobs on a single machine. Manuscript, UC Berkeley, (1997)

[GVY93a] N. Garg, V. V. Vazirani, and M. Yannakakis. Primal-dual approximation algorithms for integral flow in trees, with applications to matching and set cover. In *Proceedings of the 20th International Colloquium on Automata, Languages and Programming*, (1993)

[GVY94] N. Garg, V. V. Vazirani, and M. Yannakakis. Multiway cuts in directed and node weighted graphs. In *Proceedings of the 21th International Colloquium on Automata, Languages and Programming*, (1994) 487–498

[GT88] A. V. Goldberg and R. E. Tarjan. A new approach to the maximum flow problem. *J. of ACM*, **35** (1988) 921–940

[HO94] J. Hao and J. B. Orlin. A faster algorithm for finding the minimum cut in a graph. *Journal of Algorithms* **17** (1994) 424–446

[HN94] D. S. Hochbaum and J. Naor. Simple and fast algorithms for linear and integer programs with two variables per inequality. *SIAM Journal on Computing*, **23** (1994) 1179–1192

[HMNT93] D. S. Hochbaum, N. Megiddo, J. Naor and A. Tamir. Tight bounds and 2-approximation algorithms for integer programs with two variables per inequality. *Mathematical Programming*, **62** (1993) 69–83

[HP97] D. S. Hochbaum, A. Pathria. Approximating a Generalization of MAX 2SAT and MIN 2SAT. Manuscript, UC Berkeley, January (1997)

[Hoc82] D. S. Hochbaum. Approximation algorithms for the set covering and vertex cover problems. *SIAM J. Comput.* **11** (1982) 555-556. An extended version in: W.P. #64-79-80, GSIA, Carnegie-Mellon University, April 1980

[Hoc83] D. S. Hochbaum. Efficient bounds for the stable set, vertex cover and set packing problems. *Discrete Applied Mathematics*, **6** (1983) 243–254

[Hoc96] D. S. Hochbaum. Approximating covering and packing problems: set cover, vertex cover, independent set and related problems. Chapter 3 in *Approximation algorithms for NP-hard problems* edited by D. S. Hochbaum. PWS Boston. (1996)

[Hoc96a] D. S. Hochbaum. Various notions of approximations: Good, Better, Best and more. Chapter 9 in *Approximation algorithms for NP-hard problems* edited by D. S. Hochbaum. PWS Boston. (1996)

[Hoc97] D. S. Hochbaum. Approximating clique and biclique problems. Manuspcript, UC berkeley, June (1997). Revised Jan (1998)

[Hoc97a] D. S. Hochbaum. A framework for half integrality and good approximations. Manuspcript, UC berkeley, June 1996. Revised Sep (1997)

[Hoc98] D. S. Hochbaum. The t- vertex cover problem: Extending the half integrality framework with budget constraints. UC berkeley, manuscript. February (1998). Extended abstract in this volume.

[KKM94] R. Kohli, R. Krishnamurti and P. Mirchandani. The minimum satisfiability problem. *SIAM J. Discrete Mathematics*, **7** (1994) 275–283

[MR96] M. V. Marathe and S. S. Ravi. On approximation algorithms for the minimum satisfiability problem. Manuscript. Jan 1996. To appear *Information Processing Letters*.

[Orl93] J. Orlin. A faster strongly polynomial minimum cost flow algorithm. *Operations Research* 41 (1993) 338–350

[YC95] B. Yu and J. Cheriyan. Approximation algorithms for feasible cut and multicut problems. *Proc. Algorithms - ESA'95* P. Spirakis (ed.) LNCS 979, Springer, New York, (1995) 394–408

The t- Vertex Cover Problem: Extending the Half Integrality Framework with Budget Constraints

Dorit S. Hochbaum

Department of Industrial Engineering and Operations Research, and Walter A. Haas School of Business, University of California, Berkeley

Abstract. In earlier work we defined a class of integer programs with constraints that involve up to three variables each and showed how to derive superoptimal half integral solution to such problems. These solutions can be used under certain conditions to generate 2-approximations. Here we extend these results to problems involving budget constraints that do not conform to the structure of that class. Specifically, we address the t- vertex cover problem recently studied in the literature. In this problem the aim is to cover at least t edges in the graph with minimum weight collection of vertices that are adjacent to these edges.

The technique proposed employs a relaxation of the budget constraint and a search for optimal dual multiplier assigned to this constraint. The multipliers can be found substantially more efficiently than with approaches previously proposed that require the solution of a linear programming problem using the interior point or ellipsoid method. Instead of linear programming we use a combinatorial algorithm solving the minimum cut problem.

1 Introduction

The t- vertex cover problem is a generalization of the well known vertex cover problem. In the vertex cover problem we seek in a graph a set of nodes of minimum total weight, so that every edge in the graph has at least one endpoint in the set selected. The partial vertex cover problem, or the t- vertex cover seeks a subset of nodes that covers at least t edges in the graph at minimum total weight. For a graph $G = (V, E)$ when $t = |E|$ the t- vertex cover reduces to the ordinary vertex cover problem.

The t- vertex cover problem has been studied by Petrank [Pet94]. Recently Bshouty and Burroughs [BB98] proposed a 2-approximation algorithm for the problem with complexity of $O(|V|^{10} + |V|^7 \log B)$ for B the largest coefficient in the problem formulation.

We demonstrate here that a 2-approximation algorithm for the problem follows from the 2-approximation algorithm for the *Generalized Vertex Cover* established by Hochbaum in [Hoc96]. In that work a framework was developed

* Research supported in part by NSF award No. DMI-9713482, and by SUN Microsystems.

for generating superoptimal half integral solutions to a large class of problem formulations in strongly polynomial time. That superoptimal solution provides a bound on the optimum that is only better than the linear programming relaxation bound and is more efficient to compute. Although the t-vertex cover problem does not have the constraint structure require for the class of problems to which the technique is applicable - the so-called 2var structure - it is nevertheless possible to extend the same technique to the t-vertex cover by relaxing the one constraint that violates the 2var structure. Furthermore, the t-vertex cover problem is an illustration of extending the instances of problems with the 2var structure by adding a *budget constraint* or a fixed number of budget constraints. The 2-approximation algorithm generated by the technique is strongly polynomial and uses any combinatorial maximum flow minimum cut algorithm as subroutine.

The complexity of our 2-approximation algorithm for the t- vertex cover is $O(\log n)$ calls to an algorithm solving the minimum cut problem. Algorithms such as the push/relabel algorithm of Goldberg and Tarjan [GT88] can be used to achieve running time of $O(mn \log \frac{n^2}{m} \log n)$.

Notation. The graph $G = (V, E)$ is undirected, yet we use the notation (i, j) for an edge without implying an ordering. We use the standard notation of $n = |V|$ and $m = |E|$. For every problem instance P the notation opt(P) is used to denote the optimal solution value of P.

1.1   About the Vertex Cover Problem

The vertex cover problem was among the first to be proved NP-complete, [Kar72]. The first approximation algorithm for the unweighted problem is due to Gavril. The first 2-approximation algorithm for the weighted problem is due to Hochbaum [Hoc82]. That algorithm is based on solving the linear programming relaxation of the problem, or more precisely, the dual to the linear programming relaxation. The linear programming solution can be substituted by a solution to a certain associated minimum cut problem. That solution corresponds to a half integral solution that can be rounded up to a 2-approximate solution and has some additional strong properties (e.g. the integer components retain their value in an optimal solution).

The vertex cover problem and related approximation techniques are surveyed in [Hoc96a]. In [Hoc96] we introduced the half integrality framework for integer programming problems with up to three variables per inequality. This framework generalizes the results for the vertex cover problem. A by-product of the analysis is that any problem of this type is at least as hard to approximate as the vertex cover problem. It was conjectured in [Hoc83] that it is not possible to approximate vertex cover with a factor smaller than 2 in polynomial time unless NP=P. This conjecture is still unsettled today and it applies to all problems with the 2var structure. We sketch next the half integrality framework.

1.2 The Half Integrality Framework

Hochbaum [Hoc96] defined a class of integer programs on bounded variables with constraints that involve up to three variables each. A generic constraint in such integer program is of the form $ax + by \leq dz + c$, where the variable z appears only in that constraint. For such integer programs an algorithm delivering half integral superoptimal solutions in polynomial time was devised. For some of these problems it is possible to round the half integral solution to a 2-approximate solution. This class of integer programs contains numerous examples including the feasible cut problem, satisfiability problems, the biclique problem and the sparsest cut problem.

We call the constraint structure of problems in the framework – the 2var structure. For any problem with the 2var structure we use a construction that casts the problem as a certain minimum cut problem. The optimal solution to the cut problem corresponds to a solution that is integer multiple of half. Moreover, that half integral solution is superoptimal to the integer problem, meaning that its value is a lower bound on the integer optimum solution. For many of the problems we studied a certain rounding scheme leads to a feasible solution which at most doubles the value of the objective function. When such rounding exists the resulting integer solution is a 2-approximation.

2 The t-Vertex Cover Problem

In the t-vertex cover problem the aim is to cover at least t edges with a minimum cost collection of vertices adjacent to them. An edge is said to be covered by a vertex if the vertex is an endpoint of the edge.

The t-vertex cover problem is closely related to *Generalized Vertex Cover* in that not all edges must be covered. In *Generalized Vertex Cover* the amount of coverage is controlled via a penalty to the objective function associated with the uncovered edges:

$$\text{(Gen-VC)} \quad \begin{array}{ll} \text{Min} & \sum_{j \in V} w_j x_j + \sum_{(i,j) \in E} c_{ij} z_{ij} \\ \text{subject to} & x_i + x_j \geq 1 - z_{ij} \quad (i,j) \in E \\ & x_i, z_{ij} \text{ binary for all } i, j. \end{array}$$

The *Generalized Vertex Cover* is 2-approximable since a fractional half integral solution can always be rounded up while maintaining feasibility.

We set z_{ij} to be a binary variable that is equal to 1 if the edge (i,j) is *not* covered, and 0 if it is covered. Let x_j be a binary variable that is 1 if vertex j is in the cover. The requirement of covering at least t vertices is expressed as a "budget" of at most $m - t$ uncovered edges. The formulation of the problem is,

$$\text{(t-VC)} \quad \begin{array}{ll} \text{Min} & \sum_{j \in V} w_j x_j \\ \text{subject to} & x_i + x_j \geq 1 - z_{ij} \quad (i,j) \in E \\ & \sum_{(i,j) \in E} z_{ij} \leq m - t \\ & x_i, z_{ij} \text{ binary for all } i, j. \end{array}$$

Clearly the t-vertex cover problem does not have the 2var structure: The budget constraint, $\sum_{(i,j)\in E} z_{ij} \leq m-t$, has m variables instead of the allotted maximum of three, and each variable z_{ij} appears in two constraints – once in the budget constraint and once in the appropriate edge constraint.

One useful property of an optimal solution of t-VC is,

$$z_{ij} = \max\{1 - x_i - x_j, 0\}.$$

This property is shared also by the linear programming relaxation of the problem LP t-VC discussed in the next subsection.

Another property is that there is always an optimal solution in which the budget constraint is binding. This is obvious since if the budget constraint were not tight than it would have been always possible to increase the values of some variables z_{ij} without affecting the feasibility or the value of the objective function.

2.1 The Linear Programming Relaxation

Consider first LP t-VC – the linear programming relaxation of the t-VC problem in which the integrality requirement has been replaced by, $0 \leq x_i \leq 1$ and $0 \leq z_{ij} \leq 1$.

In our analysis the dual multiplier of the budget constraint, $\sum_{(i,j)\in E} z_{ij} \leq m - t$ plays an important role. For the problem LP t-VC, when $t > 0$ the dual multiplier is positive and the complementary slackness conditions imply that the constraint must be binding. We have thus established,

Lemma 1. *There exists an optimal solution to the linear programming relaxation* LP t-VC *in which the budget constraint is satisfied with equality.*

Thus budget constraint is satisfied with equality for a linear programming optimal solution.

For the formulation of LP t-VC the resolution of the solution is determined by the following result:

Lemma 2. *The largest value of the denominator in a basic solution to* LP t-VC *is* $2m$.

Proof: The size of the denominator is determined (via Cramer's rule) by the value of the largest (nonseparable) subdeterminant of the constraint matrix. The largest subdeterminant of the constraints that have at most two nonzero entries per constraint that are of absolute value 1 is at most 2, [HMNT93]. Adding the budget constraint can increase the absolute value of the largest subdeterminant by a factor of m at most.

□

It will be necessary in our analysis to ensure the uniqueness of the optimal solution and one-to-one relationship between the dual multiplier and an interval of values of $\sum_{(i,j)\in E} z_{ij}$. To that end we use a known theorem of linear

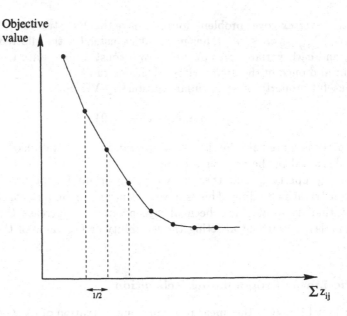

Fig. 1. The parametric linear programming relaxation objective value as a function of the number of uncovered edges

programming establishing that there exist a small enough perturbation of the objective function coefficients so that the resulting optimal solution is unique. The perturbed linear programming relaxation is:

$$\text{Min} \quad \sum_{j \in V} w_j(1 + \epsilon_j)x_j + \sum_{(i,j) \in E} \epsilon_{ij} z_{ij}$$

(LP t-VC) $\quad$ subject to $\quad x_i + x_j \geq 1 - z_{ij} \quad (i,j) \in E$

$$\sum_{(i,j) \in E} z_{ij} \leq m - t$$

$$x_i, z_{ij} \text{ binary for all } i, j.$$

The perturbation is such that $\epsilon_1 > \epsilon_2 \ldots > \epsilon_n$ and all $\epsilon_{ij} < \epsilon_n$. Arranging the edges in arbitrary order $e_1, \ldots, e_m$, one possible assignment of values is $\epsilon_i = (\frac{1}{2m})^i$ and $\epsilon_{e_j} = (\frac{1}{2m})^{n+j}$.

We will be concerned with the behavior of the optimal solution to LP t-VC as a function of $m - t$. When $t = 0$ the optimal solution value is 0. As the value of t grows (and the value of $m - t$ becomes smaller) the optimal objective value becomes larger. The optimal solution as a function of the parameter $m - t$ is piecewise linear and convex. We will investigate later the positioning of the breakpoints of this function. A description of the parametric optimal solution is given in Figure 1.

Let λ be the dual multiplier associated with the budget constraint. At an optimal solution λ^* and z_{ij}^* the complementary slackness conditions imply that,

$$\lambda^*((m - t) - \sum_{(i,j) \in E} z_{ij}^*) = 0.$$

Since $\lambda^* > 0$ for $t > 0$ it follows that the budget constraint is satisfied with equality for any optimal solution of the linear programming problem.

2.2 The Lagrangean Relaxation

We now relax the budget constraint in LP t-VC with a nonnegative Lagrange multiplier λ. The resulting relaxation for a specified value of λ is,

$$(t\text{-VC}_\lambda) \quad \begin{array}{ll} \text{Min} & \sum_{j\in V} w_j(1+\epsilon_j)x_j + \sum_{(i,j)\in E}(\lambda+\epsilon_{ij})z_{ij} - \lambda(m-t) \\ \text{subject to} & x_i + x_j \geq 1 - z_{ij} \quad (i,j)\in E \\ & 0 \leq x_i, z_{ij} \leq 1 \quad \text{for all } i,j. \end{array}$$

For any value of λ, $\text{opt}(t\text{-VC}_\lambda)\leq\text{opt}(\text{LP }t\text{-VC})$. The optimal value of t-VC$_\lambda$ is a lower bound to the optimal solution to the linear programming relaxation LP t-VC. This relaxation is also so-called *strong* Lagrangean relaxation in that there exists a value of λ for which the Lagrangean relaxation solution has the same value as the linear programming relaxation optimum. This follows from our observation that there is always an optimal solution to LP t-VC in which the budget constraint is binding.

The relaxed problem has the 2var structure and therefore has an optimal solution that is half integer. We can thus replace in the relaxation the constraint $0 \leq x_i, z_{ij} \leq 1$ for all i,j by $0 \leq x_i, z_{ij} \in \{0,\frac{1}{2},1\}$ for all i,j. We call the resulting problem $\frac{1}{2}t$-VC$_\lambda$. If the budget constraint is binding

$$\text{opt}(t - \text{VC}) \geq \text{opt}(\tfrac{1}{2}t - \text{VC}_{\lambda^*}) = \text{opt}(t - \text{VC}_{\lambda^*}) \geq \text{opt}(\text{LP}t - \text{VC}).$$

In other words, the relaxation on the half integers is only a tighter relaxation than the linear programming relaxation.

Consider again Figure 1: Every value of λ corresponds to the slope of one of the line segments in the parametric objective function. The value of the relaxation for one of these values is always attained at the rightmost end of the interval as there the objective value is the smallest.

The constant term $\lambda(m-t)$ can be omitted from the objective function. Once this term is omitted, the relaxed problem is an instance of *Generalized Vertex Cover* for any given value of λ with $c_{ij} = \lambda + \epsilon_{ij}$. For this reason it is possible to solve the relaxation in half integers more easily than the linear programming relaxation.

Lemma 3. *There exists a λ^* such that $\text{opt}(t - \text{VC}_{\lambda^*}) = \text{opt}(\text{LP } t\text{-VC})$.*

3 The 2-approximation Algorithm

The algorithm searches for smallest value of λ so that when the problem t-VC$_\lambda$ is solved for z_{ij}^* then $\sum_{(i,j)\in E} z_{ij}^* \leq m - t$. We show how to solve the relaxation in half integers so that the budget constraint is binding and demonstrate the existence of rounding of the variables to a feasible integer and 2-approximate solution.

For convenience, we use here the notation for the perturbed coefficients, $w_j' = w_j(1 + \epsilon_j)$ and $\lambda_{ij} = \lambda + \epsilon_{ij}$.

3.1 The Algorithm for Solving the Relaxation for a Given λ

The technique of "monotonizing" described in [Hoc96] is applied to the relaxation.: Each variable x_j is replaced by two variables x_j^+ and x_j^-, such that $x_j = \frac{x_j^+ - x_j^-}{2}$ with $x_j^+ \in [0,1]$ and $x_j^- \in [-1,0]$. Each variable z_{ij} is replaced by two variables z'_{ij} and z''_{ij} so that $z_{ij} = \frac{z'_{ij} + z''_{ij}}{2}$ and both z'_{ij} and z''_{ij} in $[0,1]$. The formulation of the relaxation in the new set of variables called monotonized t-VC$_\lambda$ is,

$$\begin{array}{ll} \text{Min} & \sum_{j \in V} \frac{1}{2} w'_j x_j^+ + \sum_{j \in V} \frac{1}{2} w'_j x_j^- + \frac{1}{2} \sum_{(i,j) \in E} \lambda_{ij} (z'_{ij} + z''_{ij}) \\ \text{subject to} & x_i^+ - x_j^- \geq 1 - z'_{ij} \quad (i,j) \in E \\ & -x_i^- + x_j^+ \geq 1 - z''_{ij} \quad (i,j) \in E \\ & 0 \leq x_i^+, z'_{ij}, z''_{ij} \leq 1, \quad -1 \leq x_i^- \leq 0 \quad \text{for all } i, j. \end{array}$$

To verify that this formulation is equivalent to t-VC$_\lambda$ observe that adding up the two inequalities for a given (i,j) results in $2x_i + 2x_j \geq 2 - 2z_{ij}$. Thus any solution to the monotonized formulation is feasible for the nonmonotonized formulation. The converse is true as well: given a feasible solution $\{x_j\}, \{z_{ij}\}$ to t-VC$_\lambda$, set $x_i^+ = -x_i^- = x_i$ and $z'_{ij} = z''_{ij} = z_{ij}$ for a feasible solution to monotonized t-VC. We have thus proved,

Lemma 4. *The set of feasible solutions for t-VC$_\lambda$ is identical to the set of feasible solutions to* monotonized t-VC$_\lambda$.

The formulation monotonized t-VC$_\lambda$ has a constraints' coefficients matrix that is totally unimodular. These constraints form the feasible solutions polytope of a minimum cut problem. We show how to construct a network where a minimum cut corresponds to an optimal solution to monotonized t-VC$_\lambda$.

The network has one node for each variable x_i^+ and one node for each variable x_i^-. A source node s and a sink node t are added to the network. There is an arc of capacity $\frac{1}{2} w'_i$ connecting s to each node x_i^+ and an arc of capacity $\frac{1}{2} w'_i$ connecting each node x_i^- to the sink t. For each $(i,j) \in E$ there are two arcs from node x_i^+ to the node x_j^- and another arc from x_j^+ to the node x_i^-. Both these arcs have capacities $\frac{1}{2} \lambda_{ij}$. The network is described in Figure 2.

Lemma 5. *Any finite cut separating s and t, $(S, \bar{S})$, corresponds to a feasible integer solution to* monotonized t-VC$_\lambda$.

Proof: The correspondence between the partition of nodes in the cut and the values of x_i^+ and x_i^- is set as in [Hoc96]:

$$x_i^- = \begin{cases} -1 & x_i^- \in S \\ 0 & x_i^- \in \bar{S} \end{cases}$$

$$x_i^+ = \begin{cases} 0 & x_i^+ \in S \\ 1 & x_i^+ \in \bar{S}. \end{cases}$$

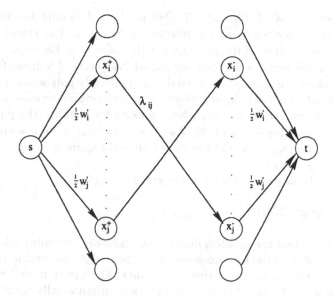

Fig. 2. The network for monotonized t-VC$_\lambda$

Let z'_{ij} be a binary variable that is equal to 1 if $x_i^+ \in S$ and $x_j^- \in \bar{S}$. Otherwise it is 0. Let $z''_{ij} = 1$ if $x_j^+ \in S$ and $x_i^- \in \bar{S}$. Otherwise it is 0. Such assignment of values creates a feasible solution to the monotonized problem. The value of a cut $(S, \bar{S})$ is,

$$
\begin{aligned}
C(S, \bar{S}) &= \tfrac{1}{2}\sum_{x_i^+ \in \bar{S}} w'_i + \tfrac{1}{2}\sum_{x_i^- \in S} w'_i + \tfrac{1}{2}\sum_{(i,j)\in E}\{\lambda_{ij} | x_i^+ \in S \text{ and } x_j^- \in \bar{S}\} \\
&\quad + \tfrac{1}{2}\sum_{(i,j)\in E}\{\lambda_{ij} | x_j^+ \in S \text{ and } x_i^- \in \bar{S}\} \\
&= \sum_i w'_i \tfrac{x_i^+ - x_i^-}{2} + \tfrac{1}{2}\sum_{(i,j)\in E}\lambda_{ij}[1 - (x_i^+ - x_j^-) + 1 - (x_j^+ - x_i^-)] \\
&= \sum_i w'_i x_i + \tfrac{1}{2}\sum_{(i,j)\in E}\lambda_{ij}(z'_{ij} + z''_{ij}) \\
&= \sum_i w'_i x_i + \sum_{(i,j)\in E}\lambda_{ij} z_{ij}.
\end{aligned}
$$

Thus the value of the cut is precisely equal to the value of the objective function of t-VC$_\lambda$. Minimizing the value of the cut minimizes also t-VC$_\lambda$. □

The minimum cut solution corresponds to an optimal integer solution to the monotonized problem. That solution in turn corresponds to a half integral optimal solution for the relaxation t-VC$_\lambda$. Rounding the half integral solution by rounding the x_j up and the z_{ij} down when fractional results in a feasible integer solution. To see that such rounding is feasible observe that whenever $z_{ij} = \tfrac{1}{2}$ then $1 - (x_i + x_j) = \tfrac{1}{2}$ and thus one of these variables is equal to $\tfrac{1}{2}$ and will be rounded up. Note that the value of $\sum_{(i,j)\in E} z_{ij}$ in the rounded solution may only go down and thus it satisfies the budget constraint.

3.2 The Search for λ^*

As the value of λ increases the value of $\sum z_{ij}$ decreases in the relaxation's optimal solution. We seek a value λ^* which is the smallest for which $\sum z_{ij} \leq m - t$. It is easy to see that $\lambda^* \in (0, \max_i w'_i]$. One method of finding the value of λ^* is to conduct binary search in the interval $(0, \max_i w'_i]$. The running time of such procedure depends on the resolution of the value of λ. Based on our earlier observation in Lemma 1 λ may assume one of $2m \cdot \max_i w'_i$ values. The binary search on such set will have polynomial (but not strongly polynomial) number of calls to the solution of the relaxed problem. Each call requires to solve a minimum cut problem which can be accomplished by using for instance the push-relabel algorithm in $O(mn \log \frac{n^2}{m})$, [GT88]. We prove now that λ may assume one of $2m$ possible values and thus the binary search will require only $\log m$ calls to a minimum cut procedure.

Let d_j be the degree of node j in the graph $G = (V, E)$.

Lemma 6. $\lambda^* \in \{\frac{(1+\epsilon_j)w'_j}{d_j} - \epsilon_{ij} | (i, j) \in E\}$.

Proof: Consider first the problem dual to the linear programming relaxation LP t-VC with λ a dual variable corresponding to the budget constraint and μ_{ij} the dual variables for the edge covering constraints. The upper bound constraints $z_{ij} \leq 1$ and $x_j \leq 1$ can be ignored as they are automatically satisfied in any optimal solution.

$$\text{(Dual } t\text{-VC)} \qquad \begin{array}{ll} \max & \sum_{(i,j) \in E} \mu_{ij} - \lambda(m - t) \\ \text{subject to} & \mu_{ij} \leq \lambda + \epsilon_{ij} \quad \text{for all } (i, j) \in E \\ & \sum_i \mu_{ij} \leq (1 + \epsilon_j)w_j \quad \text{for all } j \in V \\ & \lambda, \mu_{ij} \geq 0 \quad \text{for all } i, j. \end{array}$$

From the formulation it is evident that in an optimal solution $\{\lambda^*, \mu^*_{ij}\}$, $\lambda^* = \max_{(i,j) \in E} \mu^*_{ij} - \epsilon_{ij}$. Let $\sum_i \mu^*_{ij} = \bar{w}_j$ for all j, then $\max_i \mu^*_{ij} \geq \frac{\bar{w}_j}{d_j}$ and $\lambda^* \geq \max_{i,j}(\frac{\bar{w}_j}{d_j} - \epsilon_{ij})$. We show first that $\max_i \mu^*_{ij} = \max_j \frac{\bar{w}_j}{d_j}$ and later that $\max_i \mu^*_{ij} = \max_j \frac{w'_j}{d_j}$.

Let $\delta = \max_{(i,j) \in E} \frac{w'_j}{d_j} - \epsilon_{ij}$. We construct a feasible solution with $\lambda = \delta$ of value that is equal or larger to that of an optimal solution.

Let E^* be the set of edges (i, j) with $\mu^*_{ij} > \max_v \frac{\bar{w}_v}{d_v}$. Consider an ordering of the nodes in the graph according to the ratio, $\frac{\bar{w}_1}{d_1} < \frac{\bar{w}_2}{d_2} < \ldots < \frac{\bar{w}_n}{d_n}$. Suppose we are given an optimal dual solution with the set E^* nonempty. We will show that the values of the variables μ can be modified without affecting their sum in the objective value but permitting to reduce the value of λ and thus increasing the dual objective. This will prove that the set E^* must be empty, and therefore $\max_i \mu^*_{ij} = \max_j \frac{\bar{w}_j}{d_j}$.

Construction of a solution to Dual t-VC

Step 0: $\mu_{ij} = \min\{\mu_{ij}^*, \max_v \frac{\bar{w}_v}{d_v}\}$ for all $(i,j) \in E^*$. $k = 1$. Let $deficit(k) = \bar{w}_k - \sum_i \mu_{ik}$.

Step 1: while $deficit(k) > 0$
 begin
 For $i = k+1, \ldots n$
  if $\mu_{ik}^* < \frac{\bar{w}_k}{d_k}$ then
  $\Delta = \min\{\frac{\bar{w}_k}{d_k} - \mu_{ik}^*, deficit(k), deficit(i)\}$.
  Update $\mu_{ik} \leftarrow \mu_{ik} + \Delta$,
  $deficit(k) \leftarrow deficit(k) - \Delta$ and $deficit(i) \leftarrow deficit(k) - \Delta$.
 end.

Step 2: If $k = n$ output $\{\mu_{ij}\}_{(i,j)\in E}$, stop.
 Else set $k \leftarrow k+1$ and go to Step 1.

The procedure delivers $\sum_i \mu_{ik} = \bar{w}_k$ and every $\mu_{ij} \leq \max v \frac{\bar{w}_v}{d_v}$. Whenever Step 1 is visited there is always an edge (i,k) such that $\mu_{ik}^* < \frac{\bar{w}_k}{d_k}$. This can be verified by induction: At the beginning of iteration k, the set of edges between $\{1, \ldots, k-1\}$ and $\{k, \ldots, n\}$ has the property that the total decrease in the values of μ_{ij}^* across this cut is equal or greater to the total increase resulting from the process in step 1.

Therefore the objective function for that solution satisfies

$$\sum_{(i,j)\in E} \mu_{ij} - (\max_{j \in V} \frac{\bar{w}_j}{d_j} - \epsilon_{ij})(m - t) \geq \sum_{(i,j)\in E} \mu_{ij}^* - \lambda^*(m - t)$$

Thus we proved that in the optimal solution $\lambda^* = \max_{j \in V} \frac{\bar{w}_j}{d_j} - \epsilon_{ij}$.

We show next that there is an optimal solution so that $\max_j \frac{\bar{w}_j}{d_j} = \max_j \frac{w_j}{d_j}$.

Let $\max \frac{\bar{w}_j}{d_j} - \epsilon_{ij}$ be attained for a collection of edges $E_{\max}$ (containing possibly a single edge) of cardinality $|E_{\max}| < m - t$. Then we can reduce the value of λ^* by a small $\epsilon > 0$ by reducing all $\mu_{k,\ell}$ for edges in $E_{\max}$ by ϵ. The resulting change in the dual objective function is thus an increase of $-d_k \epsilon + (m - t)\epsilon > 0$ which contradicts the optimality of the solution.

Suppose now that $|E_{\max}| \geq m - t$, and $w_{k,\ell}' > \bar{w}_{k,\ell}$ for each $(k,\ell) \in E_{\max}$. Let $\epsilon = \min_{(k,\ell)} w_{k,\ell}' - \bar{w}_{k,\ell} > 0$. But then we can feasibly increase all $\mu_{k,\ell}^*$ by ϵ and increase also λ^* by the same amount. The resulting objective value is increased by $(|E_{\max}| - (m - t))\epsilon \geq 0$. Therefore there is an optimal solution with $\lambda^* = \max_{i,j} \frac{w_j}{d_j} - \epsilon_{ij}$ as claimed.

$\square$

We conclude that there are $2m$ breakpoints to the value of λ^* in an optimal solution.

3.3 The Approximation Algorithm

The approximation relies on properties of the parametric piecewise linear function. In particular, we show that the breakpoints are a distance of $\frac{1}{2}$ apart.

Consider the convex parametric function in Figure 1. As the value of $\sum z_{ij}$ is increasing the objective value is decreasing. The value of the dual variable λ corresponds to the (absolute value of the) slope of the function.

We sort the $2m$ breakpoints in decreasing values. For each slope (value of λ) some value of z_{ij} is increased. Because of the uniqueness, each breakpoint corresponds to a unique paring of an edge with one of its endpoints. The corresponding z_{ij} can be increased while this dual variable value is applicable. At the breakpoint the value is an integer multiple of half (since that problem's relaxation has the 2var structure). Since this slope value cannot be repeated again due to uniqueness, the value of z_{ij} will have to be incremented maximally till the next breakpoint that corresponds to a strictly lower slope of the parametric function.

2-approximation for t-VC

$L = \{\frac{w'_i}{d_j} - \epsilon_{ij}|(i,j) \in E\}$. $lower = \min_{(i,j)\in E} \frac{w'_i}{d_j} - \epsilon_{ij}$, $upper = \max \frac{w'_i}{d_j} - \epsilon_{ij}$.
begin
until $|L| = 1$ do
    $L \leftarrow L \cap [lower, upper]$.
    Select the median element λ in L.
    Call a minimum cut to solve monotonized t-VC$_\lambda$ for λ.
    If $\sum_{(i,j)\in E} z_{ij} < m - t$, then $upper = \lambda$. Else $lower = \lambda$.
    end
Round the values of x_j up, and the values of z_{ij} down. Output the solution.
end

With every call made to a minimum cut procedure the size of the set L is reduced by a factor of 2. There are thus altogether at most $\log_2(2m)$ calls. It follows that the complexity is that of $O(\log n)$ maximum flow applications.

4   General Budget Constraints

Suppose each edge has a certain benefit weight associated with covering it, a_{ij}. And suppose the problem is to cover a total of at least b weight of edges. We then substitute the budget inequality by $\sum a_{ij}(1 - z_{ij}) \geq b$. This constraint is equivalent to $\sum a_{ij}z_{ij} \leq B = \sum_{(i,j)\in E} a_{ij} - b$. We call B-VC the problem in which the budget constraint is $\sum a_{ij}z_{ij} \leq B$.

The formulation of the dual to the problem with this general requirement is,

$$\text{(Dual } B\text{-VC)} \quad \begin{array}{ll} \max & \sum_{(i,j)\in E} \mu_{ij} - \lambda B \\ \text{subject to} & \mu_{ij} \leq a_{ij}\lambda \quad \text{for all } (i,j) \in E \\ & \sum_i \mu_{ij} \leq w_j \quad \text{for all } j \in V \\ & \lambda, \mu_{ij} \geq 0 \quad \text{for all } i, j. \end{array}$$

As argued before for the case of $a_{ij} = 1$, in an optimal solution $\{\lambda^*, \mu^*_{ij}\}$ the budget constraint is binding, and $\lambda^* = \max_{(i,j)\in E} \mu^*_{ij}$.

The entire discussion is analogous: The objective function coefficients are perturbed to achieve uniqueness. The value of $\sum z_{ij}$ is increasing by a $\frac{1}{2}$ at each breakpoint. However the budget constraint value $\sum a_{ij} z_{ij}$ may increase by an arbitrary amount corresponding to some a_{ij}. Suppose the values at two consecutive breakpoints are B_1 and B_2 so that $B_1 < B < B_2$. Then we can take an appropriate convex combination of the solutions at the two breakpoints to obtain a solution with budget value B. The crucial property of this convex combination solution is that only one variable x_j and one variable z_{ij} are of value that is not integer multiple of half. By rounding all the x variable up we get a 3-approximation algorithm for the problem. Some refinements are possible and are elaborated upon in the expanded version of this paper.

This idea extends to added arbitrary number of constraints, k. The search for the λ vector becomes exponential in the number of added constraints. The approximation bound obtained is $2 + k$.

Acknowledgment

I wish to thank Lynn Burroughs for important feedback and for pointing out errors in an earlier version of this paper. My gratitude to Ilan Adler who has provided me with valuable insights on parametric linear programming.

References

[BB98] N. H. Bshouty and L. Burroughs. Massaging a linear programming solution to give a 2-approximation for a generalization of the vertex cover problem. *The Proceedings of the 15th Annual Symposium on the Theoretical Aspects of Computer Science*, (1998) 298–308

[GT88] A. V. Goldberg and R. E. Tarjan. A new approach to the maximum flow problem. *J. of ACM*, **35** (1988) 921–940

[HMNT93] D. S. Hochbaum, N. Megiddo, J. Naor and A. Tamir. Tight bounds and 2-approximation algorithms for integer programs with two variables per inequality. *Mathematical Programming*, **62** (1993) 69–83

[Hoc82] D. S. Hochbaum. Approximation algorithms for the set covering and vertex cover problems. *SIAM J. Comput.* 11 (1982) 555-556. An extended version in: W.P. #64-79-80, GSIA, Carnegie-Mellon University, April 1980.

[Hoc83] D. S. Hochbaum. Efficient bounds for the stable set, vertex cover and set packing problems. *Discrete Applied Mathematics*, 6 (1983) 243-254

[Hoc96] D. S. Hochbaum. A framework for half integrality and good approximations. Manuscript UC Berkeley, submitted. (1996). Extended abstract in this volume.

[Hoc96a] D. S. Hochbaum. Approximating covering and packing problems: set cover, vertex cover, independent set and related problems. Chapter 3 in *Approximation algorithms for NP-hard problems* edited by D. S. Hochbaum. PWS Boston (1996)

[Kar72] R. M. Karp. Reducibility among combinatorial problems. In R. E. Miller and J. W. Thatcher (eds.) *Complexity of Computer Computations*, Plenum Press, New York (1972) 85–103

[Pet94] E. Petrank. The hardness of approximation: Gap location. *Computational Complexity*, 4 (1994) 133–157

A New Fully Polynomial Approximation Scheme for the Knapsack Problem

Hans Kellerer   Ulrich Pferschy

University Graz, Department of Statistics and Operations Research
Universitätsstr. 15, A-8010 Graz, Austria
{hans.kellerer,pferschy}@kfunigraz.ac.at

Abstract. A fully polynomial approximation scheme (FPTAS) is presented for the classical 0–1 knapsack problem. The new approach considerably improves the necessary space requirements. The two best previously known approaches need $O(n + 1/\varepsilon^3)$ and $O(n \cdot 1/\varepsilon)$ space, respectively. Our new approximation scheme requires only $O(n + 1/\varepsilon^2)$ space while also reducing the running time.

1 Introduction

The classical *0–1 knapsack problem* (KP) is defined by

$$(KP) \quad \text{maximize} \quad \sum_{i=1}^{n} p_i x_i$$

$$\text{subject to} \quad \sum_{i=1}^{n} w_i x_i \leq c \qquad\qquad (1)$$

$$x_i \in \{0, 1\}, \quad i = 1, \ldots, n,$$

with p_i, w_i and c being positive integers. (Note that this integrality assumption is not necessary for the algorithm in this paper.) W.l.o.g. we assume $w_i \leq c \ \forall i = 1, \ldots, n$.

This special case of integer programming can be interpreted as filling a knapsack with a subset of the item set $\{1, \ldots, n\}$ maximizing the *profit* in the knapsack such that its *weight* is not greater than the *capacity c*.

A set of items is called *feasible* if it fulfills (1), i.e. if its total weight is at most the given capacity. The optimal solution value will be denoted by z^*.

An overview of all aspects of (KP) and its relatives is given in the book by Martello and Toth [5]. A more recent survey is given in Pisinger and Toth [6].

An algorithm A with solution value z^A is called an *ε-approximation algorithm*, $\varepsilon \in (0, 1)$, if

$$z^A \geq (1 - \varepsilon) z^*$$

holds for all problem instances. We will also call ε the *performance ratio* of A.

Basically, the developed approximation approaches can be divided into three groups:

(1) The classical *Greedy algorithm*, which is known in different versions, needs only $O(n)$ running time, requires no additional memory and has a performance ratio of $\frac{1}{2}$ (cf. also Proposition 1).

(2) *Polynomial time approximation schemes* (PTAS) reach any given performance ratio and have a running time polynomial in the length of the encoded input. The best scheme currently known is given in Caprara et al. [1] and yields a performance ratio of $\frac{1}{\ell+2}$ within $O(n^\ell)$ running time using $O(n)$ space.

(3) *Fully polynomial time approximation schemes* (FPTAS) also reach any given performance ratio and have a running time polynomial in the length of the encoded input and in the reciprocal of the performance ratio. This improvement compared to (2) is usually paid for by much larger space requirements. The best currently known FPTAS are summarized below in Table 1.

author	running time	space
Lawler [3]	$O(n \log(1/\varepsilon) + 1/\varepsilon^4)$	$O(n + 1/\varepsilon^3)$
Magazine, Oguz[4]	$O(n^2 \log n \cdot 1/\varepsilon)$	$O(n \cdot 1/\varepsilon)$
this paper	$O(n \min\{\log n, \log(1/\varepsilon)\} + 1/\varepsilon^2 \min\{n, 1/\varepsilon \log(1/\varepsilon)\})$	$O(n + 1/\varepsilon^2)$

Table 1. FPTAS for (KP)

Our contribution concerns point (3). We present an improved fully polynomial approximation scheme with running time $O(n \cdot \min\{\log n, \log(1/\varepsilon)\} + 1/\varepsilon^2 \cdot \min\{n, 1/\varepsilon \log(1/\varepsilon)\})$ and space requirement $O(n + 1/\varepsilon^2)$. In particular, the improvement in space is a major advantage for the practical use of these methods. As in [3] we will assume that arithmetic operations on numbers as large as n, c, $1/\varepsilon$ and z^* require constant time.

Our method is clearly superior to the one in [3] with a reduction by a $1/\varepsilon$ factor in space requirements. To compare the performance of our approach with the one given in [4] note that in the crucial aspect of space our new method is superior for $n \geq 1/\varepsilon$. The running time is even improved as soon as $n \log n \geq 1/\varepsilon$. These relations are rather practical assumptions because for the solution of a knapsack problem with a moderate number of items and a very high accuracy optimal solution methods could be successfully applied. Moreover, we recall a statement by Lawler [3] that "bounds are intended to emphasize asymptotic behaviour in n, rather than ε" indicating that n is considered to be of larger magnitude than $1/\varepsilon$.

For the closely related *subset sum problem*, which is a knapsack problem where the profit and weight of each item are identical, the best known FPTAS is given by Kellerer et al. [2]. It requires only $O(n + 1/\varepsilon)$ space and $O(\min\{n \cdot 1/\varepsilon, n + 1/\varepsilon^2 \log(1/\varepsilon)\})$ time. This could be seen as an improvement by a factor of $1/\varepsilon$ compared to this paper for a problem with "dimension reduced by one" although it requires different techniques.

Before going into the details of the algorithm in Section 2 we first give an informal description of our approach.

All items with "small" profits are separated in the beginning. The range of all remaining "large" profit values is partitioned into intervals of identical length. Every such interval is further partitioned into subintervals of a length increasing with the profit value, which means that the higher the profit in an interval, the fewer subintervals are generated.

Then from each subinterval a certain number of items with smallest weights is selected and all profits of these items are set equal to the lower subinterval bound. This number decreases with increasing profit value of the current interval.

Dynamic programming by profits is performed with this simplified profit structure. The approximate solution value is computed by going through the dynamic programming array and determining the best combination of an entry with a greedy solution consisting only of small profit items.

To attain the improved space complexity, we do not store the corresponding subset of items for every entry of the dynamic programming array but keep only the index of the most recently added item. To reconstruct the approximate solution set only limited backtracking through these entries is possible. Most of the solution set is computed by bipartitioning the item set and recursively solving the two resulting subproblems each dealing with only half of the original items set and a reduced range of dynamic programming.

By patching together the items of the two subproblem solutions the original solution set can be constructed. Surprisingly, this recursion can be performed without increasing the overall time and space complexities (cf. Theorem 8). A related bipartitioning scheme to save space was used by Magazine and Oguz [4]. However their approach led to an increase of time complexity.

Summarizing the main new ideas, the improvement of space is attained by storing only one item for each dynamic programming entry instead of a complete subset of items. The time improvement is derived on one hand by retriving the not stored solution set via an efficient recursive partitioning of the item set and the dynamic programming array and on the other hand by a more involved partitioning and reduction of the profit space of the items and by the use of this reduced profit structure in the dynamic programming step.

2 The new fully polynomial approximation scheme

To compute bounds for z^* let us recall the *Greedy-Heuristic G*: A feasible solution is determined by sorting the items in nonincreasing order of their profit to weight ratio and then considering all the items in that order (cf. [5]). Each item is put into the knapsack of capacity c if it fits and is never taken out again. The solution value LB, which is a lower bound for z^*, is defined as the maximum of the profit computed by this procedure and the overall highest profit of any item which is considered as a possible solution on its own. It is known that G has a performance ratio of $\frac{1}{2}$.

Proposition 1 (folklore).

$$LB \leq z^* \leq 2LB \qquad \square$$

A variant of the Greedy–Heuristic G will be used in Step 2 of the approximation scheme (AS) to add items with small profit to possible solutions consisting of items with larger profit.

For technical reasons we will introduce the *refined accuracy*

$$\tilde{\varepsilon} := \frac{1}{\lceil \frac{2}{\varepsilon} \rceil} \le \frac{\varepsilon}{2}.$$

Note that $\frac{1}{\tilde{\varepsilon}}$ and $\frac{1}{\tilde{\varepsilon}^2}$ are integers. In the following we state the new approximation scheme (AS).

Algorithm (AS) :

Step 1 Reduction of the items
Compute the above lower bound LB by running G.
Let $T := \{i \mid p_i \le LB\tilde{\varepsilon}\}$ be the set of *small items*.
Let $B := \{i \mid p_i > LB\tilde{\varepsilon}\}$ be the set of *large items*.
Partition B into $1/\tilde{\varepsilon} - 1$ intervals L_j of range $LB\tilde{\varepsilon}$ such that

$$L_j := \{i \mid jLB\tilde{\varepsilon} < p_i \le (j+1)LB\tilde{\varepsilon}\}$$

with $j = 1, \ldots, 1/\tilde{\varepsilon} - 1$.
for $j = 1, \ldots, 1/\tilde{\varepsilon} - 1$ **do**
  Partition L_j into $\lceil \frac{1}{j\tilde{\varepsilon}} \rceil - 1$ subintervals L_j^k of range $jLB\tilde{\varepsilon}^2$ such that

$$L_j^k := \{i \mid jLB\tilde{\varepsilon}(1 + (k-1)\tilde{\varepsilon}) < p_i \le jLB\tilde{\varepsilon}(1 + k\tilde{\varepsilon})\}$$

with $k = 1, \ldots, \lceil \frac{1}{j\tilde{\varepsilon}} \rceil - 1$ and the remaining (possibly smaller) subinterval

$$L_j^{\lceil \frac{1}{j\tilde{\varepsilon}} \rceil} := \{i \mid jLB\tilde{\varepsilon}(1 + (\lceil \frac{1}{j\tilde{\varepsilon}} \rceil - 1)\tilde{\varepsilon}) < p_i \le (j+1)LB\tilde{\varepsilon}\}.$$

  for $k = 1, \ldots, \lceil \frac{1}{j\tilde{\varepsilon}} \rceil$ **do**
    for all $i \in L_j^k$ set $p_i := jLB\tilde{\varepsilon}(1 + (k-1)\tilde{\varepsilon})$.
    if $(|L_j^k| > \lceil \frac{2}{j\tilde{\varepsilon}} \rceil)$ **then**
      Reduce L_j^k to the $\lceil \frac{2}{j\tilde{\varepsilon}} \rceil$ items with *minimal weight* it contains.
      Delete all other items in L_j^k.
Denote by $L \subseteq B$ the set of all remaining large items.
Step 2 Computation of the solution value
  $z^A := 0$ (*solution value*)
  $S^A := \emptyset$ (*solution item set*)
  perform **dynamic programming** $(L, 2LB)$ returning $(W[\,], R[\,])$.
  Sort the items in T in nonincreasing order of their profit to weight ratios.
  for all $j \in \{1, \ldots, 2/\tilde{\varepsilon}^2\}$ with $W[j] \le c$ **do** in nonincreasing order of $W[j]$
    Add up the items from T as long as their weight sum is not more than the remaining capacity $c - W[j]$ yielding a profit of z^T.
    $z^A := \max\{j LB\tilde{\varepsilon}^2 + z^T, z^A\}$
Let j^A denote the iteration of the last update of the solution value z^A.
Put the items from T with total profit z^T inspected in iteration j^A into S^A.

Step 3 Recursive reconstruction of the solution item set
   perform **backtracking** $(R[\,], L, j^A)$ returning (z^N) and updating S^A.
   perform **recursion** $(L, j^A LB\tilde{\varepsilon}^2 - z^N)$.

comment: The sorting of T in Step 2, which adds the $O(n \log n)$ factor
to the running time, can be avoided by iteratively finding the median of the
profit/weight ratios of the small items and bipartitioning at this point. We do
not give any details of this strategy which is analogous to Section 6 in Lawler [3]
and yields a time bound of $O(n \log(1/\varepsilon))$.

Dynamic Programming (L', P')

Input: $L' \subseteq L$: subset of items, P': profit bound.
Output: $W[\,], R[\,]$: dynamic programming arrays, $W[j] = w$ and $R[j] = r$
   means that there exists a subset of items with profit $jLB\tilde{\varepsilon}^2$, weight $w \leq c$
   and that item r was most recently added to this set.
(Note that all profit values are multiples of $LB\tilde{\varepsilon}^2$.)

$u := P'/(LB\tilde{\varepsilon}^2)$
for $j = \frac{1}{\tilde{\varepsilon}}, \frac{1}{\tilde{\varepsilon}} + 1, \ldots, u$ **do** $W[j] := c + 1$, $R[j] := 0$
$W[0] := 0$
for all distinct profit values p_t of items in L' **do**
   Let $w_1^t \leq \ldots \leq w_d^t$ be the weights of the items with profit p_t.
   for $j = 0, \frac{1}{\tilde{\varepsilon}}, \frac{1}{\tilde{\varepsilon}} + 1, \ldots, u$ **do** $label[j] :=$ false
      (indicates if entry $W[j]$ was considered before for p_t)
   $q := p_t/(LB\tilde{\varepsilon}^2)$
   for $j = 0, \frac{1}{\tilde{\varepsilon}}, \frac{1}{\tilde{\varepsilon}} + 1, \ldots, u - q$ **do**
      if $(label[j] =$ false and $W[j] + w_1^t < W[j + q])$ **then**   (2)
         $P := W[j + q]$, $W[j + q] := W[j] + w_1^t$ *(update operation)*
         $R[j + q] :=$ index of the item with weight w_1^t
         $label[j + q] :=$ true
         *(After updating an entry of $W[\,]$ we consider all possible
         updates by items with profit p_t originating from this entry)*
         if $(j + 2q \leq u)$ **then**
            $s := 2$, $\ell := 1$, $stop :=$ false
            repeat
               $P_1 := P + w_1^t$, $P_* := W[j + \ell q] + w_s^t$
               (P denotes the old value of the current entry.)
               $\ell := \ell + 1$, $P := W[j + \ell q]$
               $M := \min\{P, P_1, P_*\}$
               case $M = P$:
                  $stop :=$ true
               case $M = P_1$:
                  $W[j + \ell q] := P_1$
                  $R[j + \ell q] :=$ index of item with weight w_1^t
                  $label[j + \ell q] :=$ true, $s := 2$

128

           case $M = P_*$:

$$W[j + \ell q] := P_*$$
$$R[j + \ell q] := \text{index of item with weight } w_s^t$$
$$label[j + \ell q] := true, \; s := s + 1$$
$$\textbf{until } (stop = true \text{ or } j + (\ell + 1)q > u \text{ or } s > d)$$

return $(W[\,], R[\,])$.

comment: We perform in principle classical dynamic programming by profits. The index of the last item used to reach any profit value is stored in $R[\,]$ for further use in **backtracking**. After updating an entry $j + q$ we immediately consider the possibilities that either the old value of $W[j+q]$ plus w_1^t or the new value of $W[j+q]$ plus w_2^t yield an update of $W[j + 2q]$. If this is indeed the case the strategy is continued iteratively.

Backtracking $(R[\,], L', j^S)$

Input: $R[\,]$: dynamic programming array with the same meaning
as in **dynamic programming**.
$L' \subseteq L$: subset of items, j^S: starting point in the array.
Output: z^N: collected part of the solution value.
updates S^A.

$v := j^S, \; z^N := 0;$
repeat
    $r := R[v]$
    $S^A := S^A \cup \{r\}$
    $z^N := z^N + p_r$
    $\bar{v} := v, \; v := v - p_r/(LB\tilde{\varepsilon}^2)$
until $(v = 0 \text{ or } W[v] < W[\bar{v}] - w_r)$
return (z^N).

comment: The subset of items with total profit $j^S LB\tilde{\varepsilon}^2$ is partially reconstructed. The backtracking stops, if an entry is reached which was updated after the desired subset was constructed. Such an updated entry must not be used because it may originate from an entry with smaller profit which was generated by an item already used in the partial solution. This would result in a solution with a duplicate item.

In the following we denote by $D(L')$ the number of *distinct profit values* of items from a set L'.

Recursion $(\tilde{L}, \tilde{P})$

Input: $\tilde{L} \subseteq L$: subset of items, $\tilde{P}$: given profit value.
indirectly updates S^A by calling **backtracking**.

Partitioning
Partition $\tilde{L}$ into two disjoint subsets $\tilde{L}_1, \tilde{L}_2$ such that
$D(\tilde{L}_1) \approx D(\tilde{L}_2) \approx D(\tilde{L})/2$.
Perform **dynamic programming** $(\tilde{L}_1, \tilde{P})$ returning $(W_1[\,], R_1[\,])$.
Perform **dynamic programming** $(\tilde{L}_2, \tilde{P})$ returning $(W_2[\,], R_2[\,])$.
Merging
Find indices j_1, j_2 such that

$$(j_1 + j_2)LB\tilde{\varepsilon}^2 = \tilde{P} \quad \text{and} \quad W[j_1] + W[j_2] \text{ is minimal.} \tag{3}$$

Reconstruction and Recursion
Perform **backtracking** $(R_1[\,], \tilde{L}_1, j_1)$ returning (z_1^N).
if $(j_1 LB\tilde{\varepsilon}^2 - z_1^N > 0)$ **then**
    Perform **recursion** $(\tilde{L}_1, j_1 LB\tilde{\varepsilon}^2 - z_1^N)$.
    Perform **dynamic programming** $(\tilde{L}_2, \tilde{P})$ returning $(W_2[\,], R_2[\,])$.
    (*This recomputation is necessary because the space for $W_2[\,], R_2[\,]$*
    is used during the recursion on $\tilde{L}_1$.)
Perform **backtracking** $(R_2[\,], \tilde{L}_2, j_2)$ returning (z_2^N).
if $(j_2 LB\tilde{\varepsilon}^2 - z_2^N > 0)$ **then**
    Perform **recursion** $(\tilde{L}_2, j_2 LB\tilde{\varepsilon}^2 - z_2^N)$.

comment: Recursion does not yield an explicit return value but implicitly determines the solution set by the executions of **backtracking** during the recursion.

Concerning the practical behaviour of (AS) it can be noted that its performance hardly depends on the input data and is thus not sensitive to "hard" (KP) instances. Furthermore, it can be expected that the tree-like structure implied by **recursion** (cf. Lemma 6) will hardly be fully generated and the maximum recursion depth should be fairly moderate in practice.

3 Analysis

At first we show that the reduction of the large items B to set L in Step 1 induces only minor changes in an optimal solution. For a capacity $c' \leq c$ let z^L and z^B be the optimal solution value of the knapsack problem with item set L and item set B, respectively.

Lemma 2.

$$z^L \geq (1 - \tilde{\varepsilon})z^B$$

Proof. Let μ_j be the number of items from L_j in the optimal solution. As each item of L_j has a profit greater than $jLB\tilde{\varepsilon}$ we get immediately

$$z^B > \sum_j \mu_j jLB\tilde{\varepsilon}. \tag{4}$$

However, the rounding down of the profits in Step 1 diminishes the profit of every item in L_j by at most $jLB\tilde{\varepsilon}^2$. By bounding the new profit of the same item set as above we get

$$z^L \geq z^B - \sum_j \mu_j jLB\tilde{\varepsilon}^2 \geq z^B - \tilde{\varepsilon}z^B \geq (1 - \tilde{\varepsilon})z^B$$

by inserting (4).

The possible reduction of each set L_j^k to $\lceil \frac{2}{j\tilde{\varepsilon}} \rceil$ items with minimal weight has no effect on z^L. Because of Proposition 1 we have

$$\left\lceil \frac{2}{j\tilde{\varepsilon}} \right\rceil \cdot jLB\tilde{\varepsilon} \geq 2LB \geq z^B$$

and hence there can never be more than $\lceil \frac{2}{j\tilde{\varepsilon}} \rceil$ items from L_j in any feasible solution. Naturally, selecting from items with identical profits those with smallest weight will not decrease the optimal solution value. □

Lemma 3. Dynamic Programming (L', P') *computes for every* $P \leq P'$ *a feasible set of items from* $L' \subseteq L$ *with total profit* P *and minimal weight, if it exists.*

Proof. After the preprocessing in Step 1 all item profits are multiples of $LB\tilde{\varepsilon}^2$. As in classical dynamic programming schemes, an entry $W[j]$ contains the smallest weight of all subsets of items with a total profit of $jLB\tilde{\varepsilon}^2$ considered up to any point. The correctness is clear for each item with weight w_1^t. The other items with the same profit but larger weight are considered during the "repeat – until" loop. In this way the number of operations is proportional to the number of distinct profit values instead of the (larger) number of items (cf. Theorem 8).

For every entry $W[j]$ we have to determine the minimum over three values: (1) the previous entry in $W[j]$ (which means that there is no further update during this loop), (2) the previous entry in $W[j-q]$ plus w_1^t (a classical update by the "most efficient"item with profit p_t) and (3) the entry in $W[j-q]$ if it was recently updated by some weight w_{s-1}^t, plus w_s^t.

In this way, it is taken care that no item is used twice to achieve a particular profit value. (Note that for every entry $W[j]$, if there is no update by an item with weight w_s^t there will also be no update for an item with weight $w_{s'}^t$ for $s' > s$.) □

Lemma 4. *At the end of Step 2 we have*

$$z^A \geq (1 - \varepsilon)z^*.$$

Proof. Let us partition the optimal solution value into $z^* = z_B^* + z_T^*$, where z_B^* denotes the part contributed by large items and z_T^* the part summing up the small items in the optimal solution. The corresponding total weight of the large items will be denoted by $c_B^* \leq c$. In Lemma 2 it was shown that there exists also a set of items in L with total value at least $z_B^* - \tilde{\varepsilon}z_B^*$ and weight at most c_B^*.

Hence, it follows from Lemma 3 that the first execution of **dynamic programming** generates an entry in the dynamic programming array corresponding to a set of items with profit between $(1 - \tilde{\varepsilon})z_B^*$ and z_B^* and weight not greater than c_B^*.

In Step 2, the corresponding entry in $W[\,]$ is considered at some point during the **for**–loop and small items are added to the corresponding subset of items in a greedy way to fill as much as possible of the remaining capacity which is at least $c - c_B^*$. However, the profit difference between an optimal algorithm and the greedy heuristic in filling any capacity is less than the largest profit of an item and hence in our case it is at most $LB\tilde{\varepsilon}$. Altogether, this yields by Proposition 1 and Lemma 2

$$z^A \geq (1 - \tilde{\varepsilon})z_B^* + z_T^* - LB\tilde{\varepsilon} \geq z^* - 2\tilde{\varepsilon}\,z^* \geq (1 - \varepsilon)z^*$$

by definition of $\tilde{\varepsilon}$. □

Lemma 5. *After performing* **backtracking** $(R[\,], L', j^S)$ *a subset of items from* L' *with total profit* z^N *was put into* S^A *and there exists a subset of items in* L' *with total profit* z^R *such that*

$$z^N + z^R = j^S LB\tilde{\varepsilon}^2.$$

Proof. omitted. □

Lemma 6. *If there exists a feasible subset of* $\tilde{L}$ *with total profit* $\tilde{P}$, *then the execution of* **recursion** $(\tilde{L}, \tilde{P})$ *will add such a subset to* S^A.

Proof. Naturally, a feasible subset of items with profit $\tilde{P}$ can be divided into two parts belonging to $\tilde{L}_1$ and $\tilde{L}_2$ respectively. Thereby, the given profit value $\tilde{P}$ can be written as a sum of $\tilde{P}_1$ and $\tilde{P}_2$, each corresponding to the profit of items from one subset. Applying Lemma 3 to both subsets with profit value $P' = \tilde{P}$, it follows that in **dynamic programming** sets of items with minimal weight summing up to every reachable profit smaller or equal $\tilde{P}$ are computed. Hence, the values $\tilde{P}_1$ and $\tilde{P}_2$ will also be attained in the dynamic programming arrays and the corresponding indices j_1, j_2 fulfilling (3) can be found e.g. by going through array $W_1[\,]$ in increasing and through $W_2[\,]$ in decreasing order.

To analyze the overall effect of executing **recursion** also the recursive calls for $\tilde{L}_1$ and $\tilde{L}_2$ must be analyzed. This recursive structure of **recursion** corresponds to an ordered, *binary rooted tree* which is not necessarily complete. Each node in the tree corresponds to a call to **recursion** with the root corresponding to the first call in Step 3. A node may have up to two *child nodes*, the *left* child corresponding to the call of **recursion** with $\tilde{L}_1$ and the *right* child corresponding to a call with $\tilde{L}_2$. The order of computation corresponds to a preorder tree walk as the left child (if it exists) is always visited first. This tree model will be also used in the proof of Theorem 8.

The above statement will be shown by backwards induction moving "upwards" in the tree, i.e. starting with its leaves and applying induction to the inner nodes.

The leaves of the tree are executions of **recursion** with no further recursive calls. Hence, the two corresponding conditions yield $z_1^N + z_2^N = (j_1 + j_2)LB\tilde{\varepsilon}^2 = \tilde{P}$ and the statement of the Lemma follows from Lemma 5.

If the node under consideration has one or two childs, the corresponding calls of **recursion** follow immediately after **backtracking** on the same set of parameters. Hence, Lemma 5 guarantees that the existence condition required for the statement is fulfilled. By induction, during the processing of the child nodes items with total profit

$$j_1 LB\tilde{\varepsilon}^2 - z_1^N + j_2 LB\tilde{\varepsilon}^2 - z_2^N$$

are added to S^A. Together with the items of profit z_1^N and z_2^N added by **backtracking** this proves the above statement. $\qquad\square$

The above Lemmata can be summarized in

Theorem 7. *Algorithm* (AS) *is a fully polynomial approximation scheme for* (KP).

Proof. Lemma 4 showed that the solution value of (AS), which can be written as $z^A = z^T + j^A LB\tilde{\varepsilon}^2$, is close enough to the optimal solution value. At the end of Step 2 we put small items with profit z^T into S^A. In Step 3 we add items with profit z^N to S^A and by Lemma 5 we know the existence of an item set with profit $j^A LB\tilde{\varepsilon}^2 - z^N$. But this is precisely the condition required by Lemma 6 to guarantee that during the recursion items with a total profit of $j^A LB\tilde{\varepsilon}^2 - z^N$ are added to S^A. $\qquad\square$

It remains to analyze the asymptotic running time and space requirements of (AS).

Theorem 8. *For every performance ratio* ε, $(0 < \varepsilon < 1)$, *algorithm* (AS) *runs in time* $O(n \cdot \min\{\log n, \ \log(1/\varepsilon)\} + 1/\varepsilon^2 \cdot \min\{n, \ 1/\varepsilon \log(1/\varepsilon)\})$ *and space* $O(n + 1/\varepsilon^2)$.

Proof. The maximal number of items in L is

$$\sum_{j=1}^{1/\tilde{\varepsilon}-1} \left\lceil \frac{1}{j\tilde{\varepsilon}} \right\rceil \cdot \left\lceil \frac{2}{j\tilde{\varepsilon}} \right\rceil \approx \frac{2}{\tilde{\varepsilon}^2} \sum_{j=1}^{1/\tilde{\varepsilon}} \frac{1}{j^2}$$

which is of order $O(1/\varepsilon^2)$.

The four dynamic programming arrays clearly require $O(1/\varepsilon^2)$ space. To avoid the implicit use of memory in the recursion we always use the same memory positions for the four arrays. Therefore, $W_2[\ ]$ and $R_2[\ ]$ have to be recomputed in **recursion** because their original entries were destroyed during the recursion for $\tilde{L}_1$.

The recursive bipartitions of $D(\tilde{L})$ in **recursion** can be handled without using additional memory e.g. by partitioning $\tilde{L}$ into a set with smaller profits and one with larger profits. In this case every subset is a consecutive interval of indices and can be referenced by its endpoints.

Hence, every execution of **recursion** requires only a constant amount of additional memory. As the recursion depth is bounded by $O(\log(1/\varepsilon))$ (see below), the overall space bound follows.

The reduction of the items in every interval L_j^k can be done by a linear median algorithm. Hence, the overall running time of Step 1 is in $O(n + 1/\varepsilon^2)$.

The main computational effort is spent on **dynamic programming**. A straightforward implementation of a classical dynamic programming scheme without the complicated loop used in our procedure would consider every item for every entry of $W[\,]$ and hence require $O(|L'| \cdot P'/(LB\varepsilon^2))$ time. For the first call in Step 2 this would be $O(1/\varepsilon^4)$.

In our improved version **dynamic programming** we combine items with identical profits. For every profit value p_t each entry of the dynamic programming array is considered only a constant number of times. It is for sure considered in (2). If the label of the current entry is "false", we may enter the loop. As long as this "repeat – until" loop is continued, the label of every considered entry is set to "true". At the point when any such entry is considered again in (2) it will therefore not start another "repeat" loop.

With the above, every execution of **dynamic programming** (L', P') requires only $O(D(L') \cdot P'/(LB\varepsilon^2))$ time. This clearly dominates the effort of the following **backtracking** procedure.

After Step 1 the number of large distinct profit values is bounded by

$$D(L) \le \sum_{j=1}^{1/\tilde{\varepsilon}-1} \left\lceil \frac{1}{j\tilde{\varepsilon}} \right\rceil \approx \frac{1}{\tilde{\varepsilon}} \sum_{j=1}^{1/\tilde{\varepsilon}} \frac{1}{j} \approx \frac{1}{\tilde{\varepsilon}} \log\left(\frac{1}{\tilde{\varepsilon}}\right)$$

and hence is in $O(\min\{n, 1/\varepsilon \log(1/\varepsilon)\})$. Therefore, performing **dynamic programming** $(L, 2LB)$ in Step 2 takes $O(\min\{n, 1/\varepsilon \log(1/\varepsilon)\}1/\varepsilon^2)$ time.

The "for" loop in Step 2 can be performed by going through each $W[\,]$ and T only once after sorting $W[\,]$. As mentioned in the comment after Step 3 the $O(n \log n)$ factor caused by sorting T can be replaced by $O(n \log(1/\varepsilon))$ following an iterative median strategy by Lawler [3].

Summarizing, algorithm (AS) can be performed in $O(n \cdot \min\{\log n, \log(1/\varepsilon)\} + 1/\varepsilon^2 \cdot \min\{n, 1/\varepsilon \log(1/\varepsilon)\})$ time plus the effort of **recursion**.

To estimate the running time of **recursion** we go back to the representation of the recursive structure as a binary tree as introduced in the proof of Lemma 6. A node is said to have *level ℓ* if there are $\ell - 1$ nodes on the path to the root node. The root node has level 0. Obviously, the level of a node is equivalent to its recursion depth and gives the number of bipartitions of the initial set of distinct profit values. Therefore, the maximum level of a node is $\log D(L)$ which is in $O(\log(1/\varepsilon))$. Moreover, for a node with level ℓ we have $D(\tilde{L}) \le D(L)/2^\ell$.

The running time of a node corresponding to a call of **recursion** $(\tilde{L}, \tilde{P})$ will be interpreted as the computational effort in this procedure without the possible two recursive calls of **recursion** for $\tilde{L}_1$ and $\tilde{L}_2$. It is easy to see that the running time of a node with level ℓ is dominated by the two executions of **dynamic programming** and therefore bounded by $O((D(\tilde{L}_1)+D(\tilde{L}_2))\cdot\tilde{P}/(LB\varepsilon^2))$, which is in $O(D(L)/2^\ell \cdot \tilde{P}/(LB\varepsilon^2))$.

To find the median of $D(\tilde{L})$ appropriate auxiliary values such as the number of items with identical profit have to be kept. Without going into technical details it is clear that this can be done easily within the given time bounds.

For the combined input profit of the children in any node we get

$$j_1 LB\tilde{\varepsilon}^2 - z_1^N + j_2 LB\tilde{\varepsilon}^2 - z_2^N < (j_1+j_2)LB\tilde{\varepsilon}^2 = \tilde{P}.$$

Repeating this argument for all nodes from the root towards the leaves of the tree, this means that the sum of input profits for all nodes with equal level is less than $2LB$, the input profit of the root node. Let the number of nodes with level ℓ be $m_\ell \le 2^\ell$. If we denote the input profit of a node i in level ℓ by $\tilde{P}_\ell^i$ this means that

$$\sum_{i=1}^{m_\ell} \tilde{P}_\ell^i < 2LB.$$

Summing up over all levels the total computation time for all nodes with level ℓ this finally yields

$$\sum_{\ell=0}^{\log D(L)} \sum_{i=1}^{m_\ell} \frac{D(L)}{2^\ell} \cdot \tilde{P}_\ell^i/(LB\varepsilon^2) \le \sum_{\ell=0}^{\infty} \frac{D(L)}{2^\ell} \cdot 2/\varepsilon^2 \le 2D(L) \cdot 2/\varepsilon^2,$$

which is of order $O(1/\varepsilon^2 \cdot \min\{n, 1/\varepsilon \log(1/\varepsilon)\})$. Therefore, the recursive structure does not cause an increase in asymptotic running time and the theorem is proven. □

References

1. A. Caprara, H. Kellerer, U. Pferschy, D. Pisinger, "Approximation algorithms for knapsack problems with cardinality constraints", *Technical Report* **01/1998**, Faculty of Economics, University Graz, submitted.
2. H. Kellerer, R. Mansini, U. Pferschy, M.G. Speranza, "An efficient fully polynomial approximation scheme for the subset-sum problem", *Technical Report* **14/1997**, Fac. of Economics, Univ. Graz, submitted, see also *Proceedings of the 8th ISAAC Symposium, Springer Lecture Notes in Computer Science* **1350**, 394–403, 1997.
3. E. Lawler, "Fast approximation algorithms for knapsack problems", *Mathematics of Operations Research* **4**, 339–356, 1979.
4. M.J. Magazine, O. Oguz, "A fully polynomial approximation algorithm for the 0–1 knapsack problem", *European Journal of Operational Research*, **8**, 270–273, 1981.
5. S. Martello, P. Toth, *Knapsack Problems*, J. Wiley & Sons, 1990.
6. D. Pisinger, P. Toth, "Knapsack Problems", in D.Z. Du, P. Pardalos (eds.) *Handbook of Combinatorial Optimization*, Kluwer, Norwell, 1–89, 1997.

On the Hardness of Approximating Spanners

Guy Kortsarz

Department of Computer Science, The Open university, Klauzner 16, Ramat Aviv, Israel.

Abstract. A $k-$spanner of a connected graph $G = (V, E)$ is a subgraph G' consisting of all the vertices of V and a subset of the edges, with the additional property that the distance between any two vertices in G' is larger than that distance in G by no more than a factor of k. This paper concerns the hardness of finding spanners with the number of edges close to the optimum. It is proved that for every fixed k approximating the spanner problem is at least as hard as approximating the set cover problem
We also consider a weighted version of the spanner problem. We prove that in the case $k = 2$ the problem admits an $O(\log n)-$ratio approximation, and in the case $k \geq 5$, there is no $2^{\log^{1-\epsilon} n}-$ ratio approximation, for any $\epsilon > 0$, unless $NP \subseteq DTIME(n^{polylog\ n})$.

1 Introduction

The concept of *graph spanners* has been studied in several recent papers, in the context of communication networks, distributed computing, robotics and computational geometry [ADDJ-90, C-94, CK-94, C-86, DFS-87, DJ-89, LR-90, LR-93, LS-93, CDNS-92, PS-89, PU-89]. Consider a connected simple graph $G = (V, E)$, with $|V| = n$ vertices. A subgraph $G' = (V, E')$ of G is a $k - spanner$ if for every $u, v \in V$,

$$\frac{dist(u, v, G')}{dist(u, v, G)} \leq k,$$

where $dist(u, v, G')$ denotes the distance from u to v in G', i.e., the minimum number of edges in a path connecting them in G'. We refer to k as the *stretch factor* of G'.

In the Euclidean setting, spanners were studied in [DFS-87, DJ-89, LL-89]. Spanners for general graphs were first introduced in [PU-89], where it was shown that for every $n-$vertex hypercube there exists a 3-spanner with no more than $7n$ edges. Spanners were used in [PU-89] to construct a new type of synchronizer for an asynchronous network. Spanners are also used to construct efficient routing tables [PU-88]. For this, and other applications, it is desirable that the spanners be as *sparse* as possible, namely, have few edges. This leads to the following problem. Let $S_k(G)$ denote the minimum number of edges in a $k-$spanner for the graph G. The *sparsest k-spanner* problem involves constructing a $k-$spanner with $S_k(G)$ edges for a given graph G. In this paper we consider the question of constructing spanners with the number of edges close to $S_k(G)$. This is motivated as follows.

It is shown in [PS-89] that the problem of determining, for a given graph $G = (V, E)$ and an integer m, whether $S_2(G) \leq m$ is NP-complete. This indicates that it is unlikely to find an exact solution for the sparsest k−spanner problem even in the case $k = 2$. In [C-94] this result is extended for any integer k.

Recently, in [VRMMR-97] it is shown that the k−spanner problem is hard even in restricted cases. For example, the problem is hard even restricted to chordal graphs. The problem was known to be hard also for bipartite graphs [C-94] (see also [VRMMR-97]).

Consequently, two possible remaining courses of action for investigating the problem are establishing global bounds on $S_k(G)$ and devising approximation algorithms for the problem.

In [PS-89] it is shown that every n−vertex graph G has a polynomial time constructible $(4k+1)$−spanner with at most $O(n^{1+1/k})$ edges, or in other words, $S_{4k+1}(G) = O(n^{1+1/k})$ for every graph G. Hence in particular, A every graph G has an $O(\log n)$−spanner with $O(n)$ edges. These results are close to the best possible in general, as implied by the lower bound given in [PS-89].

The results of [PS-89] were improved and generalized in [ADDJ-90] [CDNS-92] to the weighted case, in which there are non-negative weights associated with the edges, and the distance between two vertices is the weighted distance. Specifically, it is shown in [ADDJ-90] that given an n−vertex graph and an integer $k \geq 1$, there is a polynomially constructible $(2k + 1)$−spanner G' such that $|E(G')| < n \cdot \lceil n^{\frac{1}{k}} \rceil$. It is also proven there, that the weight (sum of weights of the edges) of the constructed spanner, is $O\left(k \cdot n^{O(1/k)}\right)$ times the weight of a minimum spanning tree.

The algorithms of [ADDJ-90, PS-89] provide us with *global* upper bounds for sparse k−spanners, i.e., general bounds that hold *for every graph*. However, it may be that for specific graphs, considerably sparser spanners exist. Furthermore, the upper bounds on sparsity given by these algorithms are small (i.e., close to n) only for large values of k. It is therefore interesting to look for *approximation algorithms*, that yield near-optimal *local* bounds applying to the specific graph at hand, by exploiting its individual properties.

The only logarithmic ratio approximation algorithm known for constructing sparse spanners is for the 2−spanner problem. Specifically, in [KP-92] an $O(\log(E/V))$ approximation algorithm is given for the 2−spanner problem. That is, given a graph $G = (V, E)$, the algorithm generates a 2−spanner $G' = (V, E')$ with $|E'| = O\left(S_2(G) \cdot \log \frac{|E|}{|V|}\right)$ edges. No *small ratio* approximation algorithm is known even for the 3−spanner problem. However, it follows from the results in [ADDJ-90] that any graph admits a 3−spanner with girth (minimum length cycle) 5. Now, every graph of girth 5 has $O(n^{3/2})$ edges. This "global" result can be considered $\sqrt{n}$ "approximation" algorithm for the k−spanner problem, for $k \geq 3$. Note that this bound can not be improved in general. Consider a projective plane of order q. A projective plane of order q is a $q + 1$-regular bipartite graph with $n = q^2 + q + 1$ vertices in each size, with the additional property that every two vertices on the same side, share *exactly* one neighbor. Such a structure is known to exist, e.g., for prime q. Clearly, the girth in this graph is 6. Thus, the only 3 (and 4) spanner for the graph, is the graph itself. Furthermore, this graph contains $\theta(n^{3/2})$ edges.

In this paper, we first prove that the (unweighted) 2–spanner problem is NP–hard to approximate even when restricted to 3–colorable graphs, within $c \log n$–ratio for some constant $c < 1$. This matches the approximation ratio of $O(\log n)$ of [KP-92]. Hence the algorithm in [KP-92] is the best possible for approximating the 2–spanner problem, up to constants.

We also show that the (unweighted) k–spanner problem is hard to approximate with small ratio, even when restricted to bipartite graphs. Specifically, we prove that for every fixed integer k, $k \geq 3$ there exist a constant $c < 1$ such that it is NP–hard to approximate the k–spanner problem on bipartite graphs, within ratio $c \log n$ (the constant c depends on (the constant) k.) This result improves the NP–hardness result, for bipartite graphs [C-94], for fixed values of k. Clearly, this result implies a similar limitation on the approximability, for general graphs.

Remark: in fact we prove that for any $k = o(\log n)$ (not necessarily fixed) there exist a constant $c < 1$ such that the k–spanner has no $c \cdot \log n/k$–ratio approximation, unless $NP \subseteq DTIME(n^{O(k)})$. (In the case $k = \Omega(\log n)$, indeed, the k–spanner problem *can* be approximated within ratio $O(1)$ since, as we said before, there is always a $\log n$ spanner with $O(n)$ edges. Hence, it is mainly interesting to prove hardness results for $k = o(\log n)$.)

Finally, we define a natural new weighted version of the spanner problem. In this version, called the edge weighted k–spanner problem, each edge $e \in E$ has a positive length $l(e)$ but also a non-negative weight $w(e)$. The goal is to find a k spanner G' with low weight. Namely, the graph G' should have stretch factor k where the distances are measured according to l, and ,also, the sum of weights $w(e)$ of the edges in G', should be as small as possible. For example, in the un-weighted case $l(e) = w(e) = 1$ for every edge e. Also, in the more common weighted case, considered in [ADDJ-90, CDNS-92], for every edge e, $w(e) = l(e)$.

For the edge weighted k–spanner problem we have the following results. We consider the case where $l(e) = 1$ for every edge and w is arbitrary. For $k = 2$, this version of the problem admits a $O(\log n)$–ratio approximation. However, for every $k \geq 5$, we prove that the problem has no $2^{\log^{1-\epsilon} n}$– ratio approximation, for any $\epsilon > 0$, unless $NP \subseteq DTIME(n^{polylog\, n})$. This later result follows by a reduction from one-round two-provers interactive proof system.

We note that ours are the first results on the hardness of approximating the spanner problem.

2   Preliminaries

First, recall the following alternative definition of spanners.

Lemma 1. [PS-89] *The subgraph $G' = (V, E')$ is a $k - $ spanner of the graph $G = (V, E)$ iff $dist(u, v, G') \leq k$ for every $(v, u) \in E$.* $\square$

Thus the (un-weighted) sparsest k−spanner problem can be restated as follows: we look for a minimum subset of edges $E' \subset E$ such that every edge e that does not belong to E' lies on a cycle of length $k + 1$ or less with edges that do belong to E'. In this case we say that e is *spanned* in E' (by the remaining edges of the cycle).

In what follows we say that two (independent) sets C and D are *cliqued*, if every vertex in C is connected to every vertex in D, thus C and D induce a complete bipartite graph. We say that C and D are matched, if $|C| = |D|$ (i.e., C and D have the same size) and every vertex in C has a unique neighbor in D (that is, the two sets induce a perfect matching).

The set-cover problem: For our purpose, it is convenient to state the set-cover problem in the following way. The input for the set-cover problem consists of a bipartite graph $G(V_1, V_2, E)$, where the edges cross from V_1 to V_2 (that is, V_1 and V_2 contain no internal edges) with n vertices on each side. The goal is to find the smallest possible subset $S \subseteq V_1$, such that every vertex in V_2 has a neighbor in S.

The following result is known [RS-97]. This result followed two results by [LY-93], and [F-96] which, however, where under weaker complexity assumption.

Theorem 1. *There exist a constant $c < 1$, such that it is NP−hard to approximate the set-cover problem within ratio $c \ln n$.* □

We need the following lemma regarding a restrictive case of the set-cover problem. Consider the ρ−set-cover problem which is the set-cover problem in the case Δ, the maximum degree of any vertex in $V_1 \cup V_2$, is bounded by n^ρ for some (fixed) $0 < \rho < 1$. The usual greedy algorithm ([J-74, L-75]) gives a $\rho \cdot \ln n$−ratio approximation algorithm for the ρ−set-cover problem. On the other hand we have:

Lemma 2. *It is NP−hard to approximate the ρ−set-cover problem within $c \cdot \rho \cdot \ln n$ ratio.*

(To see this, just consider starting with a bipartite graph G with no restriction on the degrees, and taking $n^{1/\rho - 1}$ copies of G.)

In the next section we prove our hardness result for the unweighted case for $k \geq 5$. In the full paper, we prove a similar result for $k = 3$ and $k = 4$. These results show hardness of approximation on bipartite graphs. We also defer to the full paper the hardness result for $k = 2$ (this hardness result is for 3−colorable graphs).

3 A hardness result for $k \geq 5$ in the un-weighted case

In this section we consider the hardness of approximating the k−spanner problem for integer constant odd k, $k = 2t + 1$ and $t \geq 2$. The constructed graph is bipartite hence contains no odd cycles. It follows that any $2t + 2$ spanner in such a graph is a $2t + 1$ spanner as well, since the graph has no $2t + 3$ cycles.

In summary, the lower bound on the approximation for $k = 2t+1$, on bipartite graphs, will automatically imply a lower bound for the case $k = 2t + 2$.

3.1 The construction for $k \geq 5$

We start this subsection by giving an intuitive explanation for our construction. Consider the graph $G(V_1, V_2, E)$ of the set-cover problem. Suppose that we connect V_1 in a clique (i.e., complete bipartite graph) to a new set A of n vertices. Further suppose that each vertex in the set A is connected by a collection $\mathcal{P}$ of appropriate path, to every vertex of V_2. Each such path would have length exactly $2t$. Next suppose that we prove that in any spanner closed to the optimum, in any path $\mathcal{P}$ leading from a vertex $a \in A$ and $v_2 \in V_2$, the central edge is missing. Hence, in order to find an alternative path for each such missing edge. we must connect every vertex of A by a path to every vertex of V_2 "via" the vertices of V_1 (closing a cycle of length $2t + 2$ with the missing edge). Namely, each vertex of A has to be connected to each vertex of V_2, by a path of length 2, that goes trough V_1. It is easily seen, therefore, that each vertex a of A is connected in V_1 to a subset S_a that covers V_2. The number of edges needed in the spanner will therefore be roughly $n \cdot \bar{s}$, where $\bar{s}$ is the average size of all the sets S_a. Hence it is convenient for the algorithm to find a small cover S, and connect each vertex in A to S. One difficulty is to find a construction that assures that indeed the central edge will be missing in any "good" spanner

Next, we describe the construction for the k-spanner problem in the case of *constant* odd k, $k \geq 5$. Let $k = 2t + 1$, $t \geq 2$.

Let $\epsilon > 0$ be a *constant* satisfying:

$$1 - \frac{1}{2t + 1} < \epsilon < 1 - \frac{1}{2t + 2} \tag{1}$$

Let $\delta = (2t + 1)(1 - \epsilon)$. We note that by the definition of ϵ, we have that $\epsilon < \delta < 1$. Also let δ_1 be a constant satisfying:

$$\max\{\delta, 1 - \epsilon/3\} < \delta_1 < 1 \tag{2}$$

We start the construction with an instance $G(V_1, V_2, E)$ of the $(1 - \delta_1)$-set-cover problem. That is, the maximum degree in G is bounded by $n^{1-\delta_1}$. The construction is composed of two main ingredients: the *fixed part* and the *gadgets part*. The fixed part contains the graph G and the set A. We clique A and V_1 as explained above (namely, we connect each vertex $a \in A$ to each vertex $v_1 \in V_1$). Furthermore, we have two special vertices, a_1, b_1. The vertices a_1 and b_1 are joined by an edge. Then, a_1 is connected to each vertex of A, and b_1 is connected to each vertex of V_1.

Secondly, we describe the "gadgets part" of the construction. This part of the construction is intended, to connect each vertex in A to each vertex in V_2 by a path of length $2t$.

The gadget is a union of $4 \cdot \ln n \cdot n^\epsilon$, different gadgets. In the i'th gadget we have the following ingredients. The construction of the gadget involves randomization (which, nevertheless, can be easily de-randomize).

For $1 \leq i \leq 4 \ln n \cdot n^\epsilon$ do.

- Define sets $A_1^i, A_2^i, \ldots, A_t^i$, each of them of size n. The sets corresponding to different i's are disjoint.

  For each i, the set A is matched (connected in a perfect matching) to the set A_1^i. The set A_1^i is matched to the set A_2^i, and in general, for every $1 \le j \le t - 1$, the set A_j^i is matched to the set A_{j+1}^i.

- Define sets $V_{2i}^1, V_{2i}^2, \ldots, V_{2i}^{t-1}$, each of size n. The sets corresponding to different i's are disjoint.

  The sets V_2 and V_{2i}^1 are matched. Also, for each $1 \le j \le t - 2$, the sets V_{2i}^j and V_{2i}^{j+1} are matched. Call all edges of the perfect matchings (also the above ones that match A_j^i with A_{j+1}^i) "matching edges".

- Finally, for every vertex $a_t^i \in A_t^i$ and every vertex $v_{2i}^{t-1} \in V_{2i}^{t-1}$ put an edge between a_t^i and v_{2i}^{t-1}, randomly and independently, with probability $1/n^\epsilon$. Let R_i denote the collection of random edges resulting among the two sets, V_{2i}^t and A_t^i.

We note that for each vertex $a \in A$, and gadget i, there is a unique vertex $a_t^i \in A_t^i$, that is connected to the vertex a via a path that entirely goes trough the matching edges. For this reason we throughout call a_t^i a *matched copy* of a. Similarly, every vertex $v_2 \in V_2$, has a unique matched copy $v_{2i}^{t-1} \in V_{2i}^{t-1}$ in any gadget i.

It is easy to verify that the constructed graph is bipartite.

3.2 Cycles containing R_i edges

In this subsection, we show that in a sense, the R_i edges could, without loosing much, be avoided from entering a "good" spanner. It would then follow that each edge of R_i should be spanned using a path from A_t^i to V_{2i}^{t-2}, that goes trough A, V_1 and V_2. Call such path G−path. More specifically, a G−path starts at a matched copy $a_t^i \in A_t^i$ of a vertex $a \in A$. Then the G−path goes to the sets $A_{t-1}^i, A_{t-2}^i, \ldots, A_1^i$, via the matching edges. Then the path goes to a, and then to a vertex $v_1 \in G$, and then to a neighbor (in G and $\bar{G}$) $v_2 \in V_2$ of v_1. Finally, the path continues to the matched copy v_{2i}^{t-1} of v_2 via the matching edges. Note that the edge (a_t^i, v_{2i}^{t-1}) (if present) closes a cycle of length exactly $2t + 2$ with the G−path. We call such a cycle a G−cycle.

Beside the G−cycles, the other appropriate short cycles are the following:

(i) Cycles containing only edges of R_i. Call such cycles i−cycles.

(m-i) Cycles containing edges of R_i and G, and non of the vertices of A. More specifically, one can choose a vertex $v_1 \in V_1$, Choose two neighbors v_2, u_2 of v_1, in V_2, walk in parallel, using the matching edges, to the two matched copies v_{2i}^{t-1} of v_2 and u_{2i}^{t-1} of u_2, in V_{2i}^{t-1}, and finally close the cycle using a mutual neighbor $a_t^i \in A_t^i$, of v_{2i}^{t-1} and u_{2i}^{t-1}. Call such cycles $m - i$−cycles.

In the following lemmas, we bound the (expected) number of i and $m - i$ cycles.

Lemma 3. *The expected number of i cycles (summing over all of the R_i gadgets) is bounded by $\tilde{O}(n^{1+\delta})$.*

Lemma 4. *The expected number of $m - i - cycles$ is $o(n^{1+\delta_1})$.*

Hence we summarize (using the fact that by definition $n^\delta = o(n^{\delta_1})$).

Corollary 1. *The total number of expected $i-cycles$ and $m-i-cycles$ is bounded by $o(n^{1+\delta_1})$.* □

3.3 The lower bound

In this section we give the proof of the lower bound using Corollary 1. We first need the following technical lemma. This lemma states that, with high probability, all the vertices of A are connected to all the vertices of V_2, via a path that entirely goes trough the matching edges and R_i edges.

Lemma 5. *With probability at least $1 - 1/n^2$, for each vertex $a \in A$ and vertex $v_2 \in V_2$, there exist a gadget i such that the edge (a_t^i, v_{2i}^{t-1}) was included by the random choice.*

It is easily seen that one can de-randomize the construction using the method of conditional expectation in time roughly n^k. That is, it is possible, for fixed k, to construct deterministically, in polynomial time, a structure with the desired properties of Corollary 1 and Lemma 5.

Using this lemmas, we prove our two main claims. Let s^* be the size of an optimum cover in G. Note that since G is an instance of the $(1 - \delta_1)$−set-cover problem, the size of the optimum cover is at least $s^* \geq n^{\delta_1}$.

Lemma 6. *The instance $\bar{G}$ of the $2t+1-spanner$ problem, admits a $2t+1-spanner$ with no more than $2 \cdot s^* \cdot n$ edges.*

Proof. Take into the spanner the edges touching a_1, b_1 and the edges of G. The number of edges added so far is $O(n^{2-\delta_1})$. In this way we spanned the edges of the fixed part of the construction (with stretch 1 or 3 for any edge.) I.e., we took care of the edges joining A and V_1, and the edges joining V_1 and V_2 and the edges of a_1 and b_1.

Take into the spanner all the matching edges. The number of edges added here is $\tilde{O}(t \cdot n^{1+\epsilon}) = \tilde{O}(n^{1+\epsilon})$ (the last equality is valid for fixed t or even for $t = O(\log n)$ as in our case).

It only remains to span the edges of R_i. Choose a cover $S^* \subseteq V_1$ of V_2 of size s^*. Connect all the vertices of A to all the vertices of S^*. For every vertex $v_2 \in V_2$, choose a neighbor $s \in S^*$, and add the edge (s, v_2) to the spanner. It is easy to check that all the edges of R_i are spanned now, with an arbitrary alternative $G-path$ of length exactly $2t + 1$. Note that the number of edges in this spanner is $s^* \cdot n + \tilde{O}(n^{1+\epsilon}) + O(n^{2-\delta_1})$. Now, since $n^\epsilon = o(n^{\delta_1})$, $\delta_1 > 1/2$ and $s^* \geq n^{\delta_1}$, the claim follows for large enough n. □

Lemma 7. *Given a $2t + 1-$spanner $H(V, E')$ in $\bar{G}$ with no more than $l \cdot n$ edges, for some $l > 0$, there exist a (polynomially constructible) cover S of V_2 of size $2l$ or less.*

Proof. Starting with H change to a new $2t + 1-$spanner H' as follows.

First, add to H all the edges of a_1 and b_1 (if they are not already in). Similarly, add all the matching edges and the edges of G.

Now, consider the edges in $R_i \cap E'$, namely the edges of R_i, that are in the spanner. Remove every such edge (a_t^i, v_{2i}^{t-1}), joining a matched copy a_t^i of a to a matched copy v_{2i}^{t-1} of v_2. In the present situation, several of the R_i edges may be unspanned. Add for each such edge, an alternative G-path. The resulting graph H' is still a legal $2t + 1$ spanner.

Note that (beside the matching edges) for every $i-$cycle or $m - i-$cycle, we may have entered 2 additional edges to H', one joining a vertex $a \in A$ and a vertex $v_1 \in V_1$, and the other joining the vertex v_1 to a vertex $v_2 \in V_2$. These are the two relevant edges from the $G-$path. Let num denote the number of edges in the new resulting spanner H'. By Corollary 1 num is bounded above by
$$num \le l \cdot n + \tilde{\theta}(t \cdot n^{1+\epsilon}) + o(n^{1+\delta_1}) + O(n^{2-\delta_1}).$$
Note that, now, the only way to span the R_i edges is using $G-$path.

Recall that by Lemma 5, for every vertex $a \in A$ and $v_2 \in V_2$, there are two matching copies v_{2i}^{t-1} and a_t^i of v_2 and a, that are neighbors in R_i. Since the edge $e = (v_{2i}^{t-1}, a_t^i)$ is missing from H', and we need to span this edge by a $G-$path, it follows that a is connected to a neighbor $v_1 \in V_1$ of v_2. In other words, the set S_a of neighbors of a in V_1, in the spanner H', is a *cover* of V_2 in G.

One one hand, note that the number of edges num in H' is bounded below by:
$$num \ge \sum_{a \in A} |S_a| \ge n \cdot s^* \ge n^{1+\delta_1}.$$
which implies that $num \ge n^{1+\delta_1}$. Now, since $\epsilon < \delta_1$ and $\delta_1 > 1/2$, we necessarily have that $2l \ge n^{\delta_1}$, for large enough n.

On the other hand, since:
$$num \ge \sum_{a \in A} |S_a|.$$

By averaging, it turns out that there is a cover S_a of size num/n or less. Hence the size of this set S_a is bounded by $l + \tilde{O}(t \cdot n^\epsilon) + o(n^{\delta_1}) + O(n^{1-\delta_1}) \le 2l$ (the last inequality, again, follows for large enough n).

Hence we may choose S_a as the required cover. □

Now, the main theorem easily follows. For this theorem, let c be a constant such that it is $NP-$hard to approximate set-cover within ratio $c \ln n$. Then

Theorem 2. *The $k-$spanner problem, for $k \ge 5$, can not be approximated within ratio*
$$\frac{c(1 - \delta_1)}{8} \cdot \ln n,$$
unless $P = NP$.

Proof. Again, it is only necessary to prove this result for odd values, $k = 2t + 1$, $t \geq 2$ of k. Assume an algorithm $\mathcal{A}$ that has the claimed ratio. Let G be an instance of the $(1 - \delta_1)$−set-cover as described above. Let s^* be the size of the minimum cover of V_2 in G. Construct an instance $\bar{G}$ of the $2t + 1$-spanner as described. By Lemma 6, the graph $\bar{G}$ admits a spanner with $2 \cdot s^* \cdot n$ edges. The number of vertices, $\bar{n}$, in $\bar{G}$ is $\bar{\theta}(n^{1+\epsilon})$. Thus, $\ln \bar{n} < 2 \ln n$ for large enough n. By the assumption of the theorem, the algorithm $\mathcal{A}$ would produce a spanner of size less than $(c(1 - \delta_1)/8) \cdot 2 \cdot \ln n \cdot 2 \cdot s^* \cdot n = (c(1 - \delta_1)/2) \cdot \ln n \cdot s^* \cdot n$.

Let $l = (c(1-\delta_1)/2) \cdot \ln n \cdot s^*$. By Lemma 7, one derives from this construction (in polynomial time) a cover of size no larger than $c(1 - \delta_1) \cdot \ln n \cdot s^*$. This contradicts Lemma 2. □

Similar logarithmic limitation on the approximability (but with slightly different construction) follows for the cases $k = 3$ and $k = 4$, and for the case $k = 2$ on 3−colorable graphs.

4 The weighted case

In this section, we deal with the following weighted version of the spanner problem. We are given a graph G with a weight function $w(e)$ on the edges. We assume the *length* of each edge to be 1. That is, once again, in every k−spanner, a missing edge should be replaced by a path containing k edges or less. However, here we measure the quality of the spanner by its weight, namely, the sum of weight of its edges. We look for a k−spanner with minimum weight. In this section we prove an essential difference between the approximability of the cases $k = 2$, and $k \geq 5$. First, we prove that for $k = 2$, this version of the problem admits a $O(\log n)$ ratio approximation. This is done in a way similar to [KP-92]. We sketch the variant of the method of [KP-92] needed here.

We say that a vertex v 2−helps in G' an edge $e = (w, z)$ if the two edges (v, w) and (v, z) are included in G'. I.e., in G', there is an alternative path of length 2 for e, that goes trough v.

The idea is to find a vertex v that 2−helps many edges of E, using low weight. Consider each vertex $v \in G$. Let $N(v, G)$ be the graph induced in G by the neighbors $N(v)$ of v. For every neighbor z of v, put weight $w(e)$ on z in $N(v, G)$, where $e = (z, v)$. For any subset of the vertices $V' \subseteq N(V)$, let $e(V')$ denote the number of edges inside V', and let $w_v(V')$ denote the sum of weights of the vertices of V', in $N(v, G)$. We look for a vertex v and a subset $V' \subseteq N(v)$ that achieves the following minimum:

$$\min_v \left\{ \min_{V' \subseteq N(v)} \left\{ \frac{w_v(V')}{e(V')} \right\} \right\}.$$

It is important to note that the pair v, V' achieving this minimum, can be found in polynomial time using flow techniques (cf. [GGT-89]). Given v and V', one adds the edges connecting v and V', to the spanner. Note that in this way we 2–help (or span) all the edges internal to V', using low weight. This is done in iterations, until the edges are exhausted.

It follows by a proof similar to the one in [KP-92] that this greedy algorithm is an $O(\log(|E|/|V|))$–ratio approximation algorithm for the edge weighted 2–spanner problem. Details are therefore omitted.

Theorem 3. *The edge weighted 2–spanner problem, in the case $l(e) = 1$, for every edge e, admits an $O(\log(|E|/|V|))$–ratio approximation algorithm.*

However, for every $k \geq 5$, the problem has no $2^{\log^{1-\epsilon} n}$– ratio approximation, for any $\epsilon > 0$, unless $NP \subseteq DTIME(n^{polylog\,n})$. This, for example, indicates that it is unlikely that there would be any polylogarithmic ratio approximation for the edge weighted k–spanner problem, for $k \geq 5$. This result is discussed in the next subsection.

4.1 The weighted case with $k \geq 5$

In this subsection we consider the edge weighted k–spanner problem, for $k \geq 5$, in the special case where $l(e) = 1$ for every edge e. We essentially prove hardness by giving a reduction from the one-round two-provers, interactive proof system. However, for simplicity, we abstract away the relation to the interactive proof, and describe the problem we reduce from in the following simpler way. There are two versions to the problem, a maximization version and a minimization version.

We are given a bipartite graph $G(V_1, V_2, E)$. The sets V_1 and V_2 are split into a disjoint union of k sets: $V_1 = \bigcup_{i=1}^{k} A_i$ and $V_2 =_{j=1}^{k} \bigcup B_j$.

The bipartite graph, and the partition of V_1 and V_2, induce a super-graph $\mathcal{H}$ in the following way. The vertices in $\mathcal{H}$ are the sets A_i and B_j. Two sets A_i and B_j are connected by a (super) edge in $\mathcal{H}$ iff there exist $a_i \in A_i$ and $b_j \in B_j$ which are adjacent in G. For our purposes, it is convenient (and possible) to assume that that graph $\mathcal{H}$ is regular. Say that every vertex in $\mathcal{H}$ has degree d, and hence, the number of super-edges is $h = k \cdot d$.

In the maximization version, which we call Max-rep, we need to select a single "representative" vertex $a_i \in A_i$ from each subset A_i and a single vertex "representative" $b_j \in B_j$ from each B_j. We say that a super-edge (A_i, B_j) is covered if the two corresponding representatives, are neighbors in G, namely $(a_i, b_j) \in E$. The goal is to (chose a single representative from each set and) maximize the number of super-edges covered.

Let us now recall the SAT, problem, which is the decision version of the satisfiability problem. A CNF boolean formula I is given, and the question is weather there is an assignment satisfying all the clauses. The following result follows from [FL-92] and [R-95]. See also, [LY-93].

Theorem 4. *For any $0 < \epsilon < 1$, there exist a quasi-polynomial reduction of the satisfiability problem, to an instance G of Max-rep of size n, such that if I is satisfiable, there is a set of unique representatives that cover all $h = k \cdot d$ super-edges, and if the formula is not satisfiable, in the best choice of representatives, it is possible to cover no more than $h/2^{\log^{1-\epsilon} n}$ of the super-edges.* $\square$

The following easily follows from Theorem 4.

Theorem 5. *Unless* $NP \subseteq DTIME(n^{polylog\ n})$, *for any* $\epsilon > 0$ *Max-rep admits no* $2^{\log^{1-\epsilon} n}-$*ratio approximation.* $\square$

We need a slight minimization variant of Max-rep which we call Min-rep. In this case one needs to choose a minimum size subset $\mathcal{C} \subseteq V_1 \cup V_2$. Unlike the maximum version of the problem, in the minimization version of the problem, one may choose to include in $\mathcal{C}$ many vertices of each set A_i and B_j. In Min-rep it is required to cover *every* super-edge, namely that for each super-edge (A_i, B_j) there is a pair $a_i \in A_i$ and $b_j \in B_j$ that both belong to $\mathcal{C}$, such that $(a_i, b_j) \in E$.

A limitation on the approximability of Min-rep, similar to the one for Max-rep, follows easily from Theorem 4. The reduction here is rather standard (it is implicit in [LY-93]).

Theorem 6. *Unless* $NP \subseteq DTIME(n^{polylog\ n})$, *for any* $\epsilon > 0$ *Min-rep admits no* $2^{\log^{1-\epsilon} n}-$*ratio approximation algorithm.* $\square$

By reducing Min-rep to the edge weighted $k-$spanner problem, for $k \geq 5$ we get:

Theorem 7. *Unless* $NP \subseteq DTIME(n^{polylog\ n})$, *the edge weighted* $k-$*spanner problem, for* $k \geq 5$, *admits no* $2^{\log^{1-\epsilon} n}-$*ratio approximation, for any* $\epsilon > 0$, *even when restricted to bipartite graphs.*

In conclusion, we leave open the question of weather a similar "gap" in the approximability of $k = 2$, and $k \geq 5$ is valid also in the un-weighted case. The goal is either to prove evidence for such a gap, or give a logarithmic ratio approximation algorithms for fixed values of k. Also, the case of $k = 3$ and $k = 4$ deserves attention.

Acknowledgment

The author thanks Uri Feige for many helpful discussions.

References

[ADDJ-90] I. Althöfer and G. Das and D. Dobkin and D. Joseph, Generating sparse spanners for weighted graphs, *Discrete Compu. Geometry*, 9, 1993, 81-100

[B-86] B. Bollobás, Combinatorics, *Cambridge University Press*, 1986

[C-52] H. Chernoff, A Measure of Asymptotic Efficiency for Tests of Hypothesis Based on the Sum of Observations, *Ann. Math. Stat.*, 23, 1952, 493-507

[C-86] L.P. Chew, There is a Planar Graph Almost as Good as the Complete graph, *ACM Symp. on Computational Geometry*, 1994, 169-177

[C-94] L. Cai, NP-completeness of minimum spanner problems, *Discrete Applied Math*, 9, 1993, 81-100

[CDNS-92] B. Chandra and G. Das and G. Narasimhan and J. Soares, New sparseness results for graph spanners, *Proc 8th ACM Symposium on Comp. Geometry*, 1992

[CK-94] L. Cai and M. Keil, Spanners in graphs of bounded degree, *Networks*, 24, 1994, 187-194

[DFS-87] D.P. Dobkin and S.J. Friedman and K.J. Supowit, Delaunay Graphs are Almost as Good as Complete Graphs, *Proc. 31st IEEE Symp. on Foundations of Computer Science*, 1987, 20-26

[DJ-89] G. Das and D. Joseph, Which Triangulation Approximates the Complete Graph?, *Int. Symp. on Optimal Algorithms*, 1989, 168-192

[F-96] U. Feige, A threshold of ln n for approximating set cover, *Proc. 28th ACM Symp. on Theory of Computing*, 1996, 314-318

[FL-92] U. Feige L. Lova'sz, Two-provers one-round proof systems: Their power and their problems, *Proc. 24th ACM Symp. on Theory of Computing*, 733-741, 1992

[GGT-89] G. Gallo and M.D. Grigoriadis and R.E. Tarjan, A fast Parametric maximum flow algorithm and applications, *SIAM J. on Comput*, 18, 1989, 30-55

[J-74] D.S Johnson, Approximation Algorithms for Combinatorial Problems, *J. of computer and system sciences*, 9, 1974, 256-278

[KP-92] G. Kortsarz and D. Peleg, Generating Sparse 2-spanners, *J. Algorithms*, 17, 1994, 222-236

[L-75] L. Lovász, On the ratio of Integral and Fractional Covers, *Discrete Mathematics*, 13, 1975, 383-390

[LL-89] C. Levcopoulos and A. Lingas, There is a Planar Graph Almost as Good as the Complete graph and as short as minimum spanning trees, *International Symposium on Optimal Algorithms*, LNCS-401, 1989, 9-13

[LR-90] A.L. Liestman and D. Richards, Degree-Constraint Pyramid Spanners, *J. of Parallel and Distributed Computing*, 1994

[LR-93] A.L. Liestman and T.C. Shermer, Grid Spanners, *Networks*, 23, 123-133, 1993

[LS-93] A. L. Liestman and T. C. Shermer, Additive graph Spanners, *Networks*, 23, 1993, 343-364

[LY-93] C. Lund and M. Yannakakis, On the hardness of approximating minimization problems, *Proc 25'th STOC*, 1993, 286-293

[PS-89] D. Peleg and A. Schäffer, Graph Spanners, *J. of Graph Theory*, 13, 1989, 99-116

[PU-88] D. Peleg and E. Upfal, A Tradeoff between space and efficiency for routing tables, *Journal of the ACM*, 1989, 510-530

[PU-89] D. Peleg and J.D. Ullman, An optimal Synchronizer for the Hypercube, *Siam J. on Comput.*, 18, 1989, 740-747

[R-95] R. Raz, A parallel repetition theorem, *Proc 27th ACM STOC*, 1995, 447-456

[S-92] J. Soares, Approximating Euclidean distances by small degree graphs, *University of Chicago*, No. 92-05, 1992

[RS-97] R. Raz and S. Safra, A sub constant error probability low degree test, and a sub constant error probability PCP characterization of NP, STOC, 1997, 475-484.

[VRMMR-97] G. Venkatesan and U. Rotics and M.S. Madanlal and J.A. Makowsy and C. Pandu Rangan, Restrictions of Minimum Spanner Problems, 1997, Manuscript

Approximating Circular Arc Colouring and Bandwidth Allocation in All-Optical Ring Networks

Vijay Kumar

Department of ECE
Northwestern University
Evanston, IL 60208, U.S.A.
vijay@ece.nwu.edu
http://www.ece.nwu.edu/~vijay/

Abstract. We present randomized approximation algorithms for the *circular arc graph colouring problem* and for the problem of *bandwidth allocation* in all-optical ring networks. We obtain a factor-of-$(1 + 1/e + o(1))$ randomized approximation algorithm for the arc colouring problem, an improvement over the best previously known performance ratio of 5/3. For the problem of allocating bandwidth in an all-optical *WDM* (*wavelength division multiplexing*) ring network, we present a factor-of-$(1.5+1/2e+o(1))$ randomized approximation algorithm, improving upon the best previously known performance ratio of 2.

1 Introduction

The *circular arc colouring problem* is the problem of finding a minimal colouring of a set of arcs of a circle such that no two overlapping arcs share a colour. Applications include problems in network design and scheduling. There have been several investigations of the circular arc colouring problem ([17],[5]). The problem was shown to be NP-complete by Garey, Johnson, Miller and Papadimitriou in [5]. Tucker [17] reduced the problem to an integral multicommodity flow problem. For the special case of the *proper* circular arc colouring problem (a set of circular arcs is *proper* if no arc is contained in another), as an $O(n^2)$ algorithm is due to Orlin, Bonuccelli and Bovet [12]. For the general case, Tucker [17] gave a simple approximation algorithm with an approximation ratio of 2. An approximation algorithm with a performance ratio of 5/3 is due to Shih and Hsu [15].

We present a randomized approximation algorithm that achieves a performance ratio of $1 + 1/e + o(1)$ for instances where $d = \Omega(\ln n)$, where d is the minimum number of colours needed and n the number of distinct arc endpoints. Optical networks make it possible to transmit data at very high speeds, of the order of several gigabits per second. Electronic switches can not operate at such high speeds, so to enable data transmission at high speeds, it is necessary to keep the signal in optical form. Such networks are termed *all-optical networks*.

Wavelength division multiplexing (WDM) is a technology which allows for multiple signals to be carried over a link by laser beams with different wavelengths. We can think of these signals as light beams of different colours. As a signal is to be carried by the same beam of light throughout its path, a wavelength needs to be assigned to each connectivity request. To prevent interference, wavelengths must be assigned in such a way that no two paths that share a link are assigned the same wavelength. Several different network topologies have been studied, including trees, rings, trees of rings, and meshes ([14, 8, 10, 3, 7]). Rings are a very common topology: nodes in an area are usually interconnected by means of a ring network. Also, sometimes WDM networks evolve from existing fibre networks such as SONET rings. For the problem of bandwidth allocation in rings, Raghavan and Upfal [14] give an approximation algorithm within twice the optimal. We present an algorithm that has an asymptotic performance ratio of $1.5 + 1/2e + o(1)$, except when the bandwidth requirement is very small. Communication in SONET rings requires establishing point-to-point paths and the allocation of bandwidth to paths in a conflict-free manner (see [2]). The algorithmic aspect of the task is identical to that in WDM networks. Our solution, therefore, extends to this problem as well.

2 Arc Colouring and Multicommodity Flows

We are given a family F of arcs. An *overlap set* is the set of all arcs in F that contain some particular point on the circle. We will refer to the size of the largest overlap set as the *width* of F. Let $p_0, p_1, \cdots p_{n-1}$ be the n distinct endpoints of arcs in F, in clockwise order starting from some arbitrary point on the circle. An arc that runs clockwise from p_i to p_{i+1} for some i, or from p_{n-1} to p_0, is an arc of *unit length*. The *chromatic number* of F, denoted by $\gamma(F)$, is the smallest number of colours required to colour F.

Let d be the width of F. We begin by adding extra arcs of unit length to F in order to get a family of arcs of *uniform width*, that is, one in which all overlap sets are of equal cardinality. Consider a point P on the circle, between some p_i and p_{i+1}. Let there be d' arcs containing P. We add $d - d'$ arcs of unit length to F which run from p_i to p_{i+1}. Doing this for every pair of consecutive endpoints, we obtain a new family F' of arcs which is of uniform width d. This transformation helps simplify the description of our algorithm, and it is straightforward to show that $\gamma(F) = \gamma(F')$. Now suppose we were to take each arc A_i of F' that contains the point p_0, and cut it at p_0 to obtain two arcs: arc A_i^1 beginning at p_0, and arc A_i^2 terminating at p_0. The new family of arcs obtained is equivalent to a set of intervals of the real line, and can be represented as such. Let $P_0, P_1, \cdots, P_n$ be the n endpoints of intervals in such a representation, ordered from left to right. Arcs of the type A_i^1 can be represented as intervals S_i^1 beginning at P_0, while arcs of type A_i^2 can be represented as intervals S_i^2 terminating at P_n. Any other arc runs from p_i to p_j, and does not contain p_0. Such an arc A_i is represented as an interval S_i from P_i to P_j. Let I be the resulting set of intervals. I is of uniform width, that is, there are exactly d intervals passing over any point between P_0

and P_n. I can be partitioned into d unit-width sets of intervals, $I_1, I_2, \ldots, I_d$, such that S_i^1 is contained in I_i. Note that a c-colouring of I in which each S_i^1 gets the same colour as the corresponding S_i^2 is equivalent to a c-colouring of F'. We will refer to such a colouring of I as a *circular colouring*.

Consider a multicommodity network N constructed as follows. The vertices of N are labelled $x_{i,j}$, $i = 1, 2, \ldots, d$; $j = 0, 1, 2, \ldots, 2n - 1$. An interval $S_i \in I_j$ originating at P_k and terminating at P_l is represented by an edge from $x_{i,2k}$ to $x_{i,2l-1}$. If some edge terminates at $x_{i,2k-1}$ and another begins at $x_{j,2k}$, for some i, j and k, then we add an edge from $x_{i,2k-1}$ to $x_{j,2k}$. All edges have unit capacity, and an edge can only carry an integral quantity of each commodity. If a vertex has no edges incident on it, it is removed from N. The source s_i for commodity i is located at $x_{i,0}$, and the corresponding destination t_i is the vertex $x_{j,2n-1}$ such that $S_i^2 \in I_j$.

Let us refer to the set of all vertices labelled $x_{i,j}$ for some i as *row j*. We will use the term *column i* to refer to the subgraph of N induced by the set $\{x_{i,j} \mid j = 0, 1, 2, \ldots, 2n - 1\}$. We will use the term *layer i* to mean the set of edges that run between or cross over rows $2i$ and $2i + 1$. In other words, these are the edges which have exactly one end-point in the set $\{x_{j,k} \mid j \leq 2i\}$.

Note that the number of rows in the network is $2n$. We have added the "extra" n rows for the permuting of colours between columns. Recall that when at a point P some intervals terminate and new intervals begin, the new intervals get a permutation of the colours present on the old intervals.

Lemma 1. *A feasible flow of the d commodities in N is equivalent to a circular d-colouring of I.*

The proof is deferred to the full paper. The network N can be used to decide if the given family of arcs is d-colourable. We can also use a similar technique to decide if it is k-colourable, for any $k > d$. Let F'' be a family of arcs of uniform width k obtained by adding arcs of unit length to F'. It can be easily shown that

Lemma 2. F' *is k-colourable if and only if F'' is k-colourable.*

So to determine the smallest k such that F is k-colourable, we keep adding unit-length arcs to F to obtain a successively larger unit-width family F'' of arcs till we come to a point where the width k of F'' is equal to $\gamma(F'')$. In terms of the multicommodity flow network, this is equivalent to adding extra sources and sinks as well as extra edges. The extra edges help us route the original d commodities. $k - d$ columns will have to be added to the network before a feasible flow can be found. The extra commodities are easy to route. Let N' be the flow network corresponding to F''.

Lemma 3. *If commodities $1, 2, \cdots, d$ can be routed in N' from their sources to the respective destinations, then all the commodities can be routed.*

The proof is omitted for lack of space. Lemma 3 implies that we can modify N' so that it contains the same d sources and d sinks as N. The remaining sources and sinks can be removed. Henceforth, we will consider a network N' with d commodities.

2.1 Approximating the Multicommodity Flow Problem

It is relatively straightforward to set up the multicommodity flow problem as a 0-1 integer program. Soving such a program is NP-complete [4], but relaxing the integrality condition converts it into a linear programming problem which can be solved in polynomial time. We set aside the integrality condition and obtain an optimal solution. Next, we seek to use the information contained in the fractional solution to obtain a good integer solution. To do this, we use a technique called *randomized rounding* [13]. Randomized rounding involves finding a solution to the rational relaxation of an integer problem, and using information derived from that solution to obtain a provably good integer solution.

We begin with a network that has d columns, and keep adding columns till a feasible solution to the LP relaxation is obtained. Let f be a flow obtained by solving the linear program. f can be decomposed into d flows $f_1, f_2, \cdots, f_d$, one for each commodity. Each f_i can further be broken up into a set of paths $P_1, P_2, \cdots, P_p$ from the source of commodity i to its destination. To do this, consider the edge of f_i carrying the smallest amount m_j, and find a source-destination path P_j containing that edge. Associate amount m_j with the path, and subtract amount m_j from the flow carried in f_i by each edge along this path. Repeat this process till no flow remains. This process of breaking a flow into a set of paths is called *path stripping* [13]. Note that $\sum_{j=1}^{p} m_j = 1$.

In order to obtain an integer solution, we will select one path out of these p paths, and use it to route commodity i. To select a path, we cast a p-faced die where $m_1, m_2, \cdots, m_p$ are the probabilities associated with the p faces. Performing such a selection for each commodity, we obtain a set S of d paths to route the d commodities. However, these paths may not constitute a feasible solution since some edge capacity constraints may be violated. Note that in the fractional solution, an edge can carry more than one commodity. It is possible that more than one of these commodities may select a path containing this edge, since the d coin tosses to select the paths are performed independently. However, these conflicts can be resolved by adding extra columns to the network. If there are h paths in S that pass over some edge e, $h - 1$ will have to be rerouted.

Consider all the k edges in layer i of the network. There are d paths that use some of these edges. Let d_i edges out of these k edges be contained in some path, $d_i \leq d$. That means that $r_i = d - d_i$ paths will have to be rerouted. Let $r = \max\{r_i\}$. Beginning at the first layer and proceeding layer-by-layer, arbitrarily select the paths to be rerouted in case of conflicts. No more than r paths are selected in any layer. The set of paths obtained can be routed easily by adding r columns to the network. The task is analogous to r-colouring a set of line intervals which has width r.

Thus we get a 0-1 integer flow which routes all the commodities simultaneously. The network has $k + r$ columns, corresponding to a $k + r$ colouring of the family F'' of arcs. A bound on the value of $k + r$ would give us a measure of the goodness of our solution.

2.2 Algorithm Performance

Consider again the edges in layer i of the network. Let E_i be the set of such edges. d_i of these edges are selected by the d commodities in the randomized selection step. In order to minimize r, the number of columns to be added, we need to show that d_i is close to d with high probability.

In the selection step, each commodity randomly chooses a path, thereby selecting one of the k edges of E_i. This is akin to a ball being randomly placed in one of k bins. The situation can be modelled by the classical occupancy problem [11], where d balls are to be randomly and independently distributed into k bins. Let Z be the random variable representing the number of non-empty bins at the end. It can be shown that $\mathbf{E}(Z)$ is minimized when the distribution is uniform. In that case, it is a simple exercise to show that $\mathbf{E}(Z)$ is at least $d - d/e$, where e is the base of the natural logarithm.

To get a high confidence bound on r, we use a famous result in probability theory, called *Azuma's Inequality* [1]. Let $X_0, X_1, \cdots$ be a martingale sequence, and $|X_k - X_{k-1}| \leq c_k$ for all k. Azuma's Inequality says that

Theorem 1. *For all $t > 0$ and for any $\lambda > 0$,* $\Pr[|X_t - X_0| \geq \lambda] \leq 2\exp(-\frac{\lambda^2}{2\sum_{k=1}^{t} c_k^2})$.

Let X_i denote the expected value of Z after i balls have been distributed. $X_0, X_1, \cdots, X_d$ is a martingale sequence, and application of Azuma's Inequality yields

Corollary 1. $\Pr[|Z - \mathbf{E}(Z)| \geq \lambda\sqrt{d}] \leq 2\exp(-\frac{\lambda^2}{2})$

Substituting $\sqrt{4\ln n}$ for λ, we find that with probability $1 - \frac{2}{n^2}$, Z does not deviate from its expected value by more than $2\sqrt{d\ln n}$. That is, with high probability, the number of paths that need to be rerouted due to conflicts at layer i is no more than $d/e + 2\sqrt{d\ln n}$. The probability that more than this many paths need to be rerouted at *any* layer is no more than $n \cdot \frac{2}{n^2} = O(\frac{1}{n})$. So with high probability, r, the number of additional columns, is no more than $d/e + 2\sqrt{d\ln n}$.

So we have a randomized algorithm that finds a feasible integral flow using upto $k + d/e + 2\sqrt{d\ln n}$ network columns. A feasible fractional solution requires k columns, which implies that an integer solution would have at least k columns. In terms of the original arc colouring problem, we obtain a colouring that uses $k + d/e + 2\sqrt{d\ln n}$ colours, where k is a lower bound on the number of colours required by an optimal algorithm. This gives us our result:

Theorem 2. *With high probability, our algorithm can colour a family of arcs of width d using no more than $OPT + d/e + 2\sqrt{d\ln n}$ colours, where n is the number of distinct arc endpoints and OPT the number of colours required by an optimal algorithm.*

For the case where $\ln n = o(d)$, our algorithm has an asymptotic performance ratio of $1 + 1/e$. In most applications, this is the interesting range of values. On the other hand, if d is sufficiently small, it is possible to solve the problem optimally in polynomial time: for the case when $d\ln d = O(\ln n)$, a polynomial time algorithm is presented in [5].

3   The Bandwidth Assignment Problem in Optical Rings

We can represent a ring network as a circle, using points and arcs to denote nodes and paths respectively. Let n denote the number of nodes in the ring. We are given a set of communication *requests*, where each request is a (*source, destination*) pair, and *source* and *destination* are points on the circle. The task is to associate an arc with each request that connects the corresponding source and destination, and assign a colour to each arc in such a way that the set of arcs overlapping any particular point P on the circle does not contain more than one occurence of any colour.

We define an *instance* to be a collection of requests and arcs (uncoloured as yet). That is, some of the requests may have been routed. We say that a collection C of arcs is *derivable* from I if C can be obtained from I by routing the requests of I. Let $D(I)$ denote the collection of all such sets of arcs derivable from I. A *solution* for an instance I is some $C \in D(I)$ together with a valid colouring of C. An optimal solution is the one that uses the fewest colours among all solutions.

When we solve an LP relaxation of the problem, an arc may receive several colours in fractional quantities. If arc a receives quantity x of colour i, we will say that a *fraction of weight x* of a receives colour i. An undivided arc has weight 1. While working with fractional solutions, we will freely split arcs into fractions. We will also extend the notion of weight to intervals of the real line.

We will use the term *interval set* to mean a collection of intervals of the real line. The *cross-section* of an interval set S at point P is the sum of weights of all intervals of S that contain P. The *width* of an interval set S is the largest cross-section of S over all points P. In the case of a set of unsplit intervals, the width is the size of the largest clique in the corresponding interval graph.

A *conflicting pair* of arcs is a pair of arcs (a_1, a_2) such that every point on the circle is contained in at least one of (a_1, a_2), and there is some point on the circle overlapped by both a_1 and a_2. A *parallel routing* is a collection of arcs that does not contain any conflicting pairs. In the following, we examine some interesting and helpful properties of parallel routings.

Let C be a parallel routing, and S_e the set of all arcs in C that contain a link e of the ring.

Lemma 4. S_e *does not contain the whole circle.*

Proof. Assume otherwise. Let a be the arc in S_e whose clockwise endpoint is farthest from e, and let b have the farthest anticlockwise endpoint. Clearly a and b together contain the whole circle, and overlap each other over e. This means that they constitute a conflicting pair, which is not possible in a parallel routing.

As a and b together do not contain the whole circle, some link f is not contained in either:

Lemma 5. *For every link e there is another link f such that no arc of C contains both e and f.*

Each of e and f is a *complement* of the other. The removal of e and f would break the ring into two *halves*. A *complementary bisection* $CB(e, f)$ is a pair of halves created by the removal of two links e and f which are complements of each other.

Lemma 6. *If both the endpoints of an arc lie in the same half of some complementary bisection $CB(e, f)$, then in any parallel routing, that arc is contained entirely in that half.*

The following lemma lets us restrict our attention to parallel routings in the search for an optimal solution. The simple proof is omitted and can be found in the full paper.

Lemma 7. *For any instance I there is an optimal solution Z whose arcs form a parallel routing.*

Define a (c, w) *colour partition* of a family of arcs to be a partition into c families of arcs $C_1, C_2, \cdots, C_c$ each of width 1 and an interval set S of width w. The size of such a partition is defined to be $c + w$. An optimal colour partition is one of minimum size. Colour-partitioning is related to colouring. It is easy to show that

Lemma 8. *An optimal colouring of a family of arcs that uses k colours is equivalent to an optimal colour partition of size k.*

The bandwidth allocation problem can now be looked upon as the problem of routing a set of requests to minimise the size of the optimal colour partition of the resultant routing, and obtaining such a colour partition.

3.1 The Allocation Algorithm

We can route a subset of the requests in I, the given instance, in accordance with Lemmas 6 and 7. We select a link e randomly, and let us assume for the moment that we know a link f with the following property: f is the complement of e in some optimal solution Z_{opt} whose arcs constitute a parallel routing P. According to Lemma 7, such a solution must exist. Consider a request r for which there is a source-destination path p which does not include either of e and f. In Z_{opt}, such a request r must be routed over p, in accordance with Lemma 6. We route every such request r using the corresponding p. Let I' be the instance resulting from such routing. Clearly, P is still derivable from I'.

The remaining requests are the ones whose source and destination lie in different halves of $CB(e, f)$. We will refer to such requests as *crossover* requests. We set up an integer program $IP(I')$ to route the crossover requests of I' to obtain $I'' \in D(I)$ such that the size of the colour partition of I'' is the smallest among all the members of $D(I)$. Our next step is to solve $LP(I')$, the LP relaxation of $IP(I')$ to get an optimal fractional solution Z_f.

Finally, we use information from Z_f to obtain an integer solution Z_I provably close to Z_f. We try to replace the fractional quantities in Z with integer values in such a way that the resulting solution Z_I is a feasible solution to $IP(I')$ and further, with high probability the objective function value of Z_I is close to that of Z_f.

We assumed above that we know the edge f. However, this is not so. We can remedy the situation by running the algorithm for all the possible $n-1$ choices of f and taking the best of the solutions obtained, giving us a solution no worse than the one we would get if we knew the identity of f.

3.2 Integer Programming and LP Relaxation

Let $r_1, r_2, \cdots, r_l$ be the l crossover requests in I'. Each r_i can either be routed as arc b_i, which overlaps link e, or as $\bar{b}_i$, the complementary arc. Let $\{a_1, a_2, \cdots, a_m\}$ be the collection of arcs obtained by taking all the arcs of the kind b_i and $\bar{b}_i$ together with the arcs in I'. Let x_i be the indicator variable that is 1 if r_i is routed as b_i and 0 otherwise. Since all arcs b_i must have distinct colours, we can without loss of generality require that if arc b_i is selected for r_i, it must bear colour i. Otherwise, colour i will not be used. In other words, x_i is the quantity of colour i that is used. If these colours are not sufficient for all the arcs, some arcs are allowed to remain uncoloured. Let $\{a_1, a_2, \cdots, a_m\}$ be the collection of all the arcs, including arcs of the kind b_i or $\bar{b}_i$. Let $y_{i,j}$ be the indicator variable that is 1 if arc a_i gets colour j, and 0 otherwise. $y_{i,0} = 1$ if a_i is uncoloured, and 0 otherwise. To avoid colour conflicts, we require that for each link g and colour j, no more than amount x_j of colour j should be present on all the arcs that contain link g.

A feasible solution to this integer program $IP(I')$ is equivalent to a colour partition, since each colour i is present on a family of arcs of width 1, and the uncoloured arcs constitute an interval set since none of them contain link e. The objective function is therefore set up to minimize the size of this colour partition.

Consider a feasible solution F to $IP(I')$. F contains some $I'' \in D(I)$, since each request of I' has been assigned an arc. Each colour i is present on a family of arcs of width 1, and the uncoloured arcs constitute an interval set since none of them contain link e. Therefore, F contains a colour partition of I''. As the objective function is set up to minimise the size of the colour partition, an optimal solution must represent an I'' with the smallest-sized colour partition among all the members of $D(I)$.

The next step in our algorithm is to obtain an optimal solution to $LP(I')$, the LP relaxation of $IP(I')$. But before we relax the integrality condition and solve the resulting linear program, we will introduce an additional constraint: colour i is not a valid colour for $\bar{b}_i$. This is not required in the integer program since if r_i is routed as $\bar{b}_i$, x_i is 0 and colour i is not used at all. However, with the relaxation it is possible for r_i to be split between b_i and $\bar{b}_i$. We wish to ensure that b_i and $\bar{b}_i$ do not end up sharing colour i. Since the additional constraint is redundant in the integer program, it does not alter the solution space of the integer program, and hence does not change the optimal solution. We solve the linear program to get an optimal fractional solution Z_f.

Our rounding technique requires our fractional solution to satisfy the following property:

Property 1. For each i such that $x_i \neq 0$, the corresponding $\bar{b}_i$ is entirely un-coloured.

It is not possible to express this property as a linear constraint. We modify Z_f by recolouring techniques to obtain a sub-optimal fractional solution Z that satisfies it. A description of the recolouring techniques involved is deferred to the full paper. The recolouring process may cause an increase in the objective function value and a decrease in the total amount of colour used.

The last step in our bandwidth allocation algorithm is the computation of a good integer solution Z_I to the integer program $IP(I')$ formulated above. This involves rounding off the fractional quantities in the fractional solution Z to integer values in a randomized fashion.

Z resembles the solution to the multicommodity flow problem in Section 2.1 and can be looked upon as a collection of flows. Let P be a point in the middle of link e. Quantity x_i of colour i flows from P round the circle and back to P. The flow of colour i can be decomposed into paths $P_1, P_2, \cdots, P_p$, where each P_j is a family of circular arcs of width 1. An amount m_j of colour i is associated with P_j, and $\sum_1^p m_j = x_i$. We associate probability m_j with each P_j, and use a coin toss to select one of them. With probability x_i, some P_k is selected. We use the arcs of P_k to carry a unit amount of colour i. With probability $1 - x_i$ no path is selected, in which case r_i will be routed as arc $\bar{b}_i$ (and may be selected to carry some other colour). We repeat this procedure independently for each colour i. If two or more different colours select some arc a_i, we randomly pick one of them for a_i. If no colour picks a_i, it is added to the interval set of uncoloured arcs.

At the end of this procedure, all fractional quantities have been converted into 0 or 1, and the constraints are still satisfied. Let us see how far this integer solution is from the optimal.

3.3 Algorithm Performance

First of all, the objective function value of the optimal fractional solution Z_f is obviously a lower bound on that of an optimal solution to $IP(I')$. Let us compare our final integer solution Z_I with Z_f. Let z_I, z and z^* be the objective funtion values of Z_I, Z and Z_f respectively, and let $\Delta = z_I - z^*$. In the following, we try to bound Δ. Let $\Delta_1 = z_I - z$ and $\Delta_2 = z - z^*$. Then $\Delta = \Delta_1 + \Delta_2$.

As we mentioned earlier, the objective function value may increase during the recolouring process while the amount of colour used may decrease. Let $c = \sum_1^l x_i$ denote the total amount of colour in Z. Let c_f and c' be the corresponding quantities for Z_f and Z_I respectively. Our recolouring technique achieves the following bound on Δ_2, the cost of recolouring.

Lemma 9. $0 \leq z - z^* \leq c_f - c \leq c_f/2 \leq z^*/2$.

The details are deferred to the full paper. Let us now estimate Δ_1. Let S be the uncoloured interval set contained in Z and let w be its width. Let S' be the uncoloured interval set of width w' in Z_I. Let $c = \sum_1^l x_i$ denote the total amount of colour in Z. Let c' be the corresponding quantity in case of Z_I. The following bounds, due to Hoeffding [9], are useful in estimating $c' - c$. Consider a quantity X that is the sum of n independent Poisson trials, Then

Lemma 10.

$$\Pr[X - \mathbf{E}(X) \geq n\delta] \leq e^{-n\delta^2} \qquad for \ \delta > 0. \tag{1}$$

$$\Pr[X - \mathbf{E}(X) \geq \delta \mathbf{E}(X)] \leq e^{-\frac{\delta^2}{3}\mathbf{E}(X)} \qquad where \ 0 < \delta < 1. \tag{2}$$

Let there be m colours in use in the fractional solution Z. Each of these is involved in coin-tossing and path selection.

Lemma 11. $\Pr[c' - c \geq \sqrt{m \ln c}] \leq \frac{1}{c}$

Proof. We can regard c' as a sum of m Poisson trials with associated probabilities x_i. The expected value of c' is c. The result follows from (1) when $\sqrt{\frac{\ln c}{m}}$ is substituted for δ.

Next, let us bound $|w' - w|$. We use the following result due to McDiarmid [9].

Lemma 12. *Let $X_1, X_2, \ldots, X_n$ be independent random variables, with X_i taking values in a set A_i for each i. Suppose that the (measurable) function $f :$ $\Pi A_i \to \mathcal{R}$ satisfies*
$$|f(\bar{X}) - f(\bar{X}')| \leq c_i$$
whenever the vectors $\bar{X}$ and $\bar{X}'$ differ only in the ith coordinate. Let Y be the random variable $f(X_1, X_2, \ldots, X_n)$. Then for any $\phi > 0$,
$$\Pr[|Y - \mathbf{E}(Y)| > \sqrt{\phi \sum_i c_i^2 / 2}] \leq 2e^{-\phi}.$$

Lemma 13. $\Pr[|w' - \mathbf{E}[w']| > \sqrt{m \ln c}] \leq \frac{2}{c}.$

Proof. w' is a function of m inependent choices. c_i is 1 if $x_i = 1$, and 2 if $0 < x_i < 1$. So $\sum_i c_i^2 / 2$ is no more than m, and Lemma 12 yields
$$\Pr[|w' - \mathbf{E}[w']| > \sqrt{\phi \, m}] \leq 2e^{-\phi}.$$
Substituting $\ln c$ for ϕ gives us the desired expression.

Lemmas 14 to 20 seek to bound $\mathbf{E}[w']$. Let $w_p, w_{1,p}$ and $w_{2,p}$ denote the cross-sections of S, S_1 and S_2 respectively at a point p on the ring. Let $w'_p, w'_{1,p}$ and $w'_{2,p}$ be the respective quantities for S', S'_1 and S'_2. Using Lemma 12 it is straightforward to establish the following two bounds.

Lemma 14. $\Pr[|w'_{1,p} - w_{1,p}| > \sqrt{m(\ln n + 2\ln c)}] \leq \frac{2}{nc^2}.$

Lemma 15. $\Pr[|w'_{2,p} - w_{2,p}| > c/e + \sqrt{\frac{m}{2}(\ln n + 2\ln c)}] \leq \frac{2}{nc^2}.$

Lemma 16. $\Pr[|w'_p - w_p| > c/e + (1 + 1/\sqrt{2})\sqrt{m(\ln n + 2\ln c)}] \leq \frac{4}{nc^2}.$

Proof. As $|w'_p - w_p| \leq |w'_{1,p} - w_{1,p}| + |w'_{2,p} - w_{2,p}|$, the result follows directly from Lemmas 14 and 15.

Lemma 16 implies that

Lemma 17. $\Pr[\max_p \{w'_p - w_p\} > c/e + (1 + 1/\sqrt{2})\sqrt{m(\ln n + 2\ln c)}] \le \frac{4}{c^2}.$

We will use the following lemma, the validity of which follows from the definition of expectation.

Lemma 18. *For any random variable X and value x_0, suppose that $\Pr[X \ge x_0] \le p$. Let x_{max} be the largest possible value of X. Then:*
$$\mathbf{E}(X) \le (1 - p)x_0 + px_{max}.$$

Lemma 19. $\mathbf{E}(\max_p \{w'_p - w_p\}) \le c/e + 2\sqrt{m(\ln n + \ln c)} + 8/c.$

Proof. Follows from Lemmas 17 and 18 and the fact that $w'_p - w_p$ can not exceed $2c$.

Lemma 20. $\mathbf{E}(w') \le w + c/e + 2\sqrt{m(\ln n + 2\ln c)} + 8/c.$

Proof. $w' \le w + \max_p \{w'_p - w_p\}$, so
$$\mathbf{E}(w') \le w + \mathbf{E}(\max_p \{w'_p - w_p\}).$$
This in conjunction with Lemma 19 yields the result.

The following bound on w' follows from Lemmas 13 and 20:

Lemma 21. *With probability at least $1 - \frac{2}{c}$,*
$$w' \le w + c/e + 2\sqrt{m(\ln n + 2\ln c)} + 8/c + \sqrt{m \ln c}.$$

Lemma 22. *With probability at least $1 - \frac{3}{c}$, $\Delta_1 \le c/e + 2\sqrt{m(\ln n + 2\ln c)} + 8/c + \sqrt{m \ln c}$.*

Proof. We know that $\Delta_1 = z_I - z = (c' + w') - (c + w) = (c' - c) + (w' - w)$.

Lemma 11 tells us that with probability at least $1 - \frac{1}{c}$, $c' - c < \sqrt{m \ln c}$. Lemma 21 gives us a similar high probability bound on $w' - w$. Together, they directly imply the above bound on Δ_1.

Lemma 23. *With high probability, z_I is no more than $z^*(\frac{3}{2} + \frac{1}{2e} + o(1)) + O(\sqrt{z^* \ln n})$.*

Proof. $\Delta = \Delta_1 + \Delta_2$. Let $c = c_f - x \cdot z^*$, where c_f is the total amount of colour in Z_f. This means that $c \le c_f - x \cdot c_f$, since c_f is a lower bound on z^*. Lemma 9 implies that $\Delta_2 \le x \cdot z^*$, and that x can not be more than $\frac{1}{2}$.

Together with Lemma 22, this implies that with high probability,
$$\Delta \le x \cdot z^* + c/e + 2\sqrt{m(\ln n + 2\ln c)} + 8/c + \sqrt{m \ln c}.$$
The expression on the right reduces to $z^*(x - \frac{x}{e} + \frac{1}{e} + o(1)) + O(\sqrt{z^* \ln n})$. Since $\Delta = z_I - z^*$, and x can not exceed $\frac{1}{2}$ (Lemma 9), we have our result.

z_I is the size of the colour partition of I' computed by our algorithm. Since a colour partition of size z_I results in a colouring that uses z_I colours, and since z^*, the value of an optimal solution to $LP(I')$, is a lower bound on the value of an optimal solution to $IP(I')$, we have our main result:

Theorem 3. *With high probability, our algorithm requires no more than $OPT(1.5 + 1/2e + o(1)) + O(\sqrt{OPT \ln n})$ wavelengths to accomodate a given set I of communication requests, where OPT is the smallest number of wavelengths sufficient for I.*

Some of the latest WDM systems involve over a hundred wavelengths. In such a system, $\sqrt{OPT \ln n}$ is likely to be considerably smaller than OPT, giving us a performance ratio close to $1.5 + 1/2e$. In the case of SONET rings the available bandwidth is often much larger, which means that $\ln n$ is typically $o(OPT)$.

References

1. Azuma, K. Weighted Sum of Certain Dependent Random Variables. *Tohoku Mathematical Journal*, 19:357-367, 1967.
2. Cosares, S., Carpenter, T., and Saniee, I. Static Routing and Slotting of Demand in SONET Rings. Presented at the *TIMS/ORSA Joint National Meeting*, Boston, MA, 1994.
3. Erlebach, T. and Jansen, K. Call Scheduling in Trees, Rings and Meshes. In *Proc. 30th Hawaii International Conf. on System Sciences*, 1997.
4. Even, S., Itai, A., and Shamir, A. On the complexity of timetable and multicommodity flow problems, *SIAM Journal of Computing*, 5(1976),691-703.
5. Garey, M.R., Johnson, D.S., Miller, G.L. and Papadimitriou, C.H. The Complexity of Coloring Circular Arcs and Chords. *SIAM J. Alg. Disc. Meth.*, 1(2):216-227, 1980.
6. Golumbic, M. C., *Algorithmic Graph Theory and Perfect Graphs*, Academic Press, 1980.
7. Kaklamanis, C. and Persiano, P. Efficient wavelength routing on directed fiber trees. In *Proc. 4th Annual European Symposium on Algorithms*, 1996.
8. Kumar, V. and Schwabe, E.J. Improved access to optical bandwidth in trees. In *Proc 8th Annual ACM-SIAM Symp. on Discrete Algorithms*, pp. 437-444, 1997.
9. McDiarmid, C. On the method of bounded differences. In J. Siemons, editor, *Surveys in Combinatorics*, volume 141 of *LMS Lecture Notes Series*, pages 148–188. 1989.
10. Mihail, M., Kaklamanis, C., and Rao, S. Efficient Access to Optical Bandwidth. In *Proc IEEE Symp on Foundations of Comp Sci*, pp. 548-557, 1995.
11. Motwani, R., and Raghavan, P. *Randomized Algorithms*, Cambridge University Press, 1995.
12. Orlin, J.B., Bonuccelli, M.A., and Bovet, D.P. An $O(n^2)$ Algorithm for Coloring Proper Circular Arc Graphs. *SIAM J. Alg. Disc. Meth.*, 2(2):88-93, 1981.
13. Raghavan, P. Randomized Rounding and Discrete Ham-Sandwiches: Provably Good Algorithms for Routing and Packing Problems. *PhD Thesis, CS Division, UC Berkeley*, 1986.
14. Raghavan, P., and Upfal, E. Efficient Routing in All-Optical Networks. In *Proc 26th ACM Symp on Theory of Computing*, pp. 134-143, 1994.
15. Shih, W.K. and Hsu, W.L. An Approximation Algorithm for Coloring Circular-Arc Graphs. *SIAM Conference on Discrete Mathematics, 1990*.
16. Settembre, M. and Matera, F. All-optical implementations of high capacity TDMA networks. *Fiber and Integrated Optics*, 12:173-186, 1993.
17. Tucker, A. Coloring a Family of Circular Arcs. *SIAM J. Appl. Math.*, 29(3):493-502, 1975.

Approximating Maximum Independent Set in k-Clique-Free Graphs

Ingo Schiermeyer

Lehrstuhl für Diskrete Mathematik
und Grundlagen der Informatik,
Technische Universität Cottbus, D-03013 Cottbus, Germany,
Fax: +49 355 693042
e-mail: schierme@math.tu-cottbus.de

Abstract. In this paper we study lower bounds and approximation algorithms for the independence number $\alpha(G)$ in k-clique-free graphs G. Ajtai et al. [1] showed that there exists an absolute constant c_1 such that for any k-clique-free graph G on n vertices and with average degree $\bar{d}$, $\alpha(G) \geq c_1 \frac{\log((\log \bar{d})/k)}{\bar{d}} n$.
We improve this lower bound for $\alpha(G)$ as follows: Let G be a connected k-clique-free graph on n vertices with maximum degree $\Delta(G) \leq n - 2$. Then $\alpha(G) \geq n(\bar{d}(k-2)^2 \log(\bar{d}(k-2)^2) - \bar{d}(k-2)^2 + 1)/(\bar{d}(k-2)^2 - 1)^2$ for $\bar{d} \geq 2$.
For graphs with moderate maximum degree Halldórsson and J. Radhakrishnan [9] presented an algorithm with a $O(\Delta/\log\log\Delta)$ performance ratio. We will show that $\log\log\Delta$ in the denominator can be replaced by $\log\Delta$ to improve the performance ratio of this algorithm. This is based on our improved lower bound for $\alpha(G)$ in k-clique-free graphs.
For graphs with moderate to large values of Δ Halldórsson and J. Radhakrishnan [9] presented an algorithm with a $\Delta/6(1+o(1))$ performance ratio. We will show that, for a given integer $q \geq 3$, this performance ratio can be improved to $\Delta/2q(1+o(1))$.

Key words: Graph, Maximum Independent Set, k-Clique, Algorithm, Complexity, Approximation

1 Introduction

We use Bondy & Murty [5] and Garey & Johnson [7] for terminology and notation not defined here and consider simple graphs only.

An *independent set* (*clique*) I in a graph G on n vertices is a set of vertices in which no (every) two are adjacent. The cardinality of an independent set (clique) of maximum cardinality will be denoted by $\alpha(G)$ and $\omega(G)$, respectively. Obviously, an independent set I in G forms a clique in G^C and vice versa.

The computation of $\alpha(G)$ and $\omega(G)$ are well-known NP-complete problems (cf. [7]). Considerable progress has been achieved in the last few years concerning the non-approximability of these problems.

In [2] Arora et al. proved the following theorem.

Theorem 1. *If $P \neq NP$ then there exists an $\epsilon > 0$ such that no polynomial time approximation algorithm for $\alpha(G)$ can have a performance ratio of $O(n^\epsilon)$.*

This was improved by Håstad [8] as follows.

Theorem 2. *For any $\epsilon > 0$ there is no polynomial time approximation algorithm for $\alpha(G)$ with a performance ratio of $O(n^{1-\epsilon})$ unless $NP = co - RP$.*

Boppana and Halldórsson [4] developed a Ramsey-Algorithm with a performance ratio of $O(n/log^2 n)$. In spite of considerable effort, no approximation algorithm with a better performance ratio is known so far.

Since the general problem is apparently hard, it is natural to ask what kind of restrictions make the problem easier to approximate. A large number of approximation results have been obtained in terms of the maximum vertex degree $\Delta(G)$ of G.

Among these graphs with bounded maximum degree have been studied intensively (cf. e.g. [3], [9], [10]).

For graphs with moderate maximum degree Halldórsson and J. Radhakrishnan [9] presented an algorithm with a $O(\Delta/loglog\Delta)$ performance ratio. We will show that $\log \log \Delta$ in the denominator can be replaced by $\log \Delta$ to improve the performance ratio of this algorithm. This is based on an improved lower bound for $\alpha(G)$ in k-clique-free graphs which we are going to prove in the next section.

For graphs with moderate to large values of Δ Halldórsson and J. Radhakrishnan [9] presented an algorithm with a $\Delta/6(1 + o(1))$ performance ratio. We will show that, for a given integer $q \geq 3$, this performance ratio can be improved to $\Delta/2q(1 + o(1))$.

Additional notation For an independent set algorithm *Alg*, *Alg(G)* is the size of the solution obtained by the algorithm applied on the graph G. The performance ratio of the algorithm in question is defined by

$$\rho = \max_G \frac{\alpha(G)}{Alg(G)}.$$

2   The Independence Number in k-Clique-Free Graphs

In [1] Ajtai et al. showed the following theorem for k-clique-free graphs.

Theorem 3. *There exists an absolute constant c_1 such that for any k-clique-free graph G,*

$$\alpha(G) \geq c_1 \frac{\log((\log \bar{d})/k)}{\bar{d}} n.$$

Here we show how the approach of Shearer [13], [14] can be extended to k-clique-free graphs to improve the above theorem. As a key observation we will make use of the following corollary which can be deduced from Turán's theorem [16]. For a vertex v of G let $|E(G[N(v)])|$ denote the cardinality of the edge set of $G[N(v)]$, i.e. the graph induced by the neighbours of the vertex v.

Corollary 4. *In a k-clique-free graph G,*

$$|E(G[N(v)])| \leq (d(v))^2 \frac{k-3}{2(k-2)}$$

for every vertex $v \in V(G)$.

In [13] Shearer showed for triangle-free graphs that $\alpha(G) \geq n f(\bar{d})$ where $f(d) = (d \log d - d + 1)/(d-1)^2, f(0) = 1, f(1) = 1/2$ and $\bar{d}$ is the average degree of G. Note that f is the solution to the diffential equation

$$(d+1)f(d) = 1 + (d - d^2)f'(d), \qquad f(0) = 1. \tag{1}$$

In [14] Shearer slightly strengthened this result by replacing the differential equation (1) with the difference equation

$$(d+1)f(d) = 1 + (d - d^2)[f(d) - f(d-1)], \qquad f(0) = 1. \tag{2}$$

and the term $n f(\bar{d})$ with $\sum_{i=1}^{n} f(d_i)$, where $d_1, d_2, \ldots, d_n$ is the degree sequence of the graph G.

The solution of the difference equation lies above the solution of the differential equation (for $d \geq 2$); however, for $d \to \infty$ the asymptotic behaviour is the same.

For a suitable constant c, $0 \leq c \leq 1$, we extend the difference equation (2) to the following difference equation

$$(d+1)f(d) = 1 + c(d - d^2)[f(d) - f(d-1)], \quad f(d) = \frac{1}{d+1}, 0 \leq d \leq k-2, \tag{3}$$

and will show that $\alpha(G) \geq \sum_{i=1}^{n} f(d_i)$ in k-clique-free graphs.

The corresponding differential equation is

$$(d+1)f(d) = 1 + c(d - d^2)f'(d), \qquad f(d) = \frac{1}{d+1}, 0 \leq d \leq k-2. \tag{4}$$

with the solution $f(d) = (d/c \log d/c - d/c + 1)/(d/c - 1)^2, f(d) = \frac{1}{d+1}, 0 \leq d \leq k-2$. Again the solution of the difference equation lies above the solution of the differential equation (for $d \geq 2$); however, for $d \to \infty$ the asymptotic behaviour is the same.

We will need the following technical lemma.

Lemma 5. *Suppose* $f(d) = \frac{1+c(d^2-d)f(d-1)}{cd^2+(1-c)d+1}$ *for* $d \geq 1$ *and* $f(d) = \frac{1}{d+1}, 0 \leq d \leq k-2$. *Then*

(i) $f(d) \geq \frac{1}{d+1}$ *for* $d \geq 0$,
(ii) $f(d)$ *is decreasing and*
(iii) $f(d-1) - f(d)$ *is decreasing (i.e. f is convex).*

Proof. (extract) (i) can be shown by induction using the difference equation. By induction and using the difference equation and (i) we obtain (ii). By a tedious calculation we can show that $f(d-1) - f(d+1) < 2(f(d-1) - f(d))$ implying (iii). □

Before stating our main theorem we will make some useful assumptions. Since the computation of $\alpha(G)$ is additive with respect to the components of G we may assume that G is connected. Next we may assume that G satisfies $\Delta(G) \leq n-2$. Otherwise, $\alpha(G) = \alpha(G - v)$ for a vertex $v \in V(G)$ with $d(v) = \Delta(G) = n-1$ and we could reduce the problem size.

Therefore we may assume from now on that G is connected and has maximum degree $\Delta(G) \leq n-2$. This leads to the following useful corollary.

Corollary 6. *For every vertex* $v \in V(G)$ *there are two vertices* $u \in N(v)$ *and* $w \notin N[v] = N(v) \cup \{v\}$ *such that* $uw \in E(G)$.

Theorem 7. *Let G be a connected k-clique-free graph on n vertices with* $\Delta(G) \leq n-2$. *Let* $f(d) = \frac{1}{d+1}, 0 \leq d \leq k-2, f(d) = \frac{1+c(d^2-d)f(d-1)}{cd^2+(1-c)d+1}$ *for* $d \geq k-1$. *Then* $\alpha(G) \geq \sum_{i=1}^{n} f(d_i)$ *for* $c = \frac{1}{(k-2)^2}$.

Corollary 8. *Let G be a connected k-clique-free graph on n vertices with* $\Delta(G) \leq n-2$. *Then* $\alpha(G) \geq n \cdot f(\bar{d}) \geq n \cdot f(\Delta) \geq n(\Delta(k-2)^2 \log(\Delta(k-2)^2) - \Delta(k-2)^2 + 1)/(\Delta(k-2)^2 - 1)^2$ *for* $\bar{d} \geq 2$.

Remark: If necessary, we will use in the following the notation $f_k(d)$ instead of $f(d)$.

Proof. Note that $f(d)$ and $[f(d) - f(d+1)]$ are decreasing as $d \to \infty$. We will prove the theorem by induction on n. Clearly it holds for $n = 1$. Let $S = \sum_{i=1}^{n} f(d_i)$ and let i be a vertex of G. Define N^i to be the set of neighbours of vertex i and N_2^i to be the set of vertices in G which are at distance 2 from i. For $q \in N_2^i$ let n_q^i be the number of neighbours in N^i, i.e. n_q^i is the number of common neighbours of q and i. Define H_i to be the graph formed from G by deleting i and its neighbours. Let $d_1^i, \ldots, d_{n'}^i$ be the degree sequence of H_i, where $n' = |V(H_i)|$. Let $T_i = \sum_l f(d_l^i)$. Then

$$T_i = S - f(d_i) - \sum_{j \in N^i} f(d_j) + \sum_{q \in N_2^i} [f(d_q - n_q^i) - f(d_q)]. \tag{5}$$

Hence if we can find an i such that

$$1 - f(d_i) - \sum_{j \in N^i} f(d_j) + \sum_{q \in N_2^i} [f(d_q - n_q^i) - f(d_q)] \geq 0 \qquad (6)$$

the result will follow by induction (adjoin i to a maximum independent set in H_i.) In fact we will show that (6) holds on average. Let

$$A = \sum_{i=1}^{n} [1 - f(d_i) - \sum_{j \in N^i} f(d_j) + \sum_{q \in N_2^i} [f(d_q - n_q^i) - f(d_q)]]. \qquad (7)$$

By interchanging the order of summation, we obtain

$$A = \sum_{i=1}^{n} [1 - (d_i + 1)f(d_i) + \sum_{q \in N_2^i} [f(d_i - n_q^i) - f(d_i)]]. \qquad (8)$$

Note that $q \in N_2^i \Leftrightarrow i \in N_2^q$ and $n_q^i = n_i^q$. Let

$$B_i = \sum_{q \in N_2^i} [f(d_i - n_q^i) - f(d_i)]]. \qquad (9)$$

Thus we have $f(d_i - n_q^i) - f(d_i) \geq n_q^i [f(d_i - 1) - f(d_i)]$ since $f(d) - f(d+1)$ is a decreasing function of d. Let $t_j(i)$ denote the number of neighbours of a vertex $j \in N^i$ in N^i. Hence

$$B_i \geq [\sum_{j \in N^i} (d_j - t_j(i) - 1)][f(d_i - 1) - f(d_i)]]. \qquad (10)$$

Let E be the edge set of G. Then summing up both sides of (10) over i gives

$$\sum_{i=1}^{n} B_i \geq \sum_{(i,j) \in E} [(d_j - t_j(i) - 1)[f(d_i - 1) - f(d_i)]$$
$$+ (d_i - t_i(j) - 1)[f(d_j - 1) - f(d_j)]]. \qquad (11)$$

Since $f(d) - f(d+1)$ is decreasing we have $(d_i - d_j)[[f(d_j - 1) - f(d_j)] - [f(d_i - 1) - f(d_i)]] \geq 0$ which implies

$$\sum_{i=1}^{n} B_i \geq \sum_{(i,j) \in E} [(d_i - t_j(i) - 1)[f(d_i - 1) - f(d_i)]$$
$$+ (d_j - t_i(j) - 1)[f(d_j - 1) - f(d_j)]] \qquad (12)$$

$$\geq \sum_{i=1}^{n} \max(1, (\frac{1}{k-2}d_i^2 - d_i))[f(d_i - 1) - f(d_i)]. \tag{13}$$

applying corollary 4 and 6. Especially, every missing edge in $G[N^i]$ contributes (at least) two edges between $G[N^i]$ and $G[N_2^i]$. This leads to the term $\frac{1}{k-2}d_i^2 - d_i$.

Now we try to find a suitable constant c_k such that

$$\frac{1}{k-2}d_i^2 - d_i \geq c_k(d_i^2 - d_i). \tag{14}$$

Therefore,

$$c_k \leq \frac{(1 - \frac{k-3}{k-2})d_i - 1}{d_i - 1} = 1 - \frac{k-3}{k-2} \cdot \frac{d_i}{d_i - 1}$$

and thus $c_k = \frac{1}{(k-2)^2}$ satisfies

$$\max(1, \frac{1}{k-2}d_i^2 - d_i)) \geq \frac{1}{(k-2)^2}(d_i^2 - d_i). \tag{15}$$

Substituting (13) into (8) then gives

$$A \geq \sum_{i=1}^{n} 1 - (d_i + 1)f(d_i) + c_k(d_i^2 - d_i)[f(d_i - 1) - f(d_i)]. \tag{16}$$

Hence $A \geq 0$ since we have chosen f so that each term in the sum in (16) is 0. Hence an i satisfying (6) must exist completing the proof.

$\square$

Remark: As indicated after (6) the proof provides a polynomial time algorithm to construct an independent set of size at least $\sum_{i=1}^{n} f(d_i)$. We will call this algorithm *k-clique-free Algorithm(G)*.

3 Approximating Maximum Independent Set

In [9] Halldórsson and J. Radhakrishnan present an approximation algorithm based on Theorem 3 by Ajtai, Erdös, Komlós and Szemerédi [1] named AEKS. It contains the algorithm called CliqueCollection which is based on the subgraph removal approach introduced in [4]. For a given integer l it finds in G a maximal collection of disjoint cliques of size l; in other words, S is a set of mutually nonintersecting cliques of size l such that the graph $G - S$ contains no l-cliques. Such a collection can be found in $O(\Delta^{l-1}n)$ time by exhaustive search for a (l-1)-clique in the neighbourhood of each vertex. That is polynomial whenever $l = O(\log_\Delta n)$.

As in [9] an independent set is *maximal* (MIS) if adding any further vertices to the set violates its independence. An MIS is easy to find and provides a sufficient general lower bound of $\alpha(G) \geq n/(\Delta + 1)$.

Combining CliqueCollection and MIS leads to the algorithm AEKS-SR(G) presented in [9].

AEKS-SR(G)
$G' \leftarrow G - CliqueCollection(G, c_1 \log\log \Delta)$
return max (AEKS(G'), MIS(G))
end

Theorem 9. *The performance ratio of AEKS-SR is $O(\Delta/\log\log \Delta)$.*

Using Theorem 7 we can show the following improved performance ratio of AEKS-SR.

Theorem 10. *The performance ratio of AEKS-SR is $O(\Delta/\log \Delta)$.*

Proof. Let k denote $c_1 \log \Delta$, and let n' denote the order of $V(G')$. A maximum independent set collects at most one vertex from each k-clique, for at most

$$\alpha(G) \leq n/k + n' \leq 2\max(n/k, n'),$$

while the size of the solution found by AEKS-SR is at least

$$AEKS - SR(G) \geq \max(\frac{1}{\Delta+1}n, \frac{k}{\Delta}n') \geq \frac{k}{\Delta+1} \max(n/k, n').$$

The ratio between the two of them clearly satisfies the claim. $\square$

Observe that this combined method runs in polynomial time for Δ as large as $n^{1/\log n}$.

We now turn our attention to approximation algorithms for moderate to large maximum degree. Using Theorem 7 we will show how the method presented in [9] with an asymptotic $\Delta/6(1 + o(1))$ performance ratio can be improved to achieve an asymptotic performance ratio of $\Delta/2q(1 + o(1))$ for a given integer $q \geq 3$. First we recall the used approximation algorithms.

2-opt. Khanna et al. [11] studied a simple local search algorithm called 2-opt in [9]. Starting with an initial maximal independent set I, it tries all possible ways of adding two vertices and removing only one while retaining the independence property. Hence it suffices to look at pairs adjacent to a common vertex in I. The following was shown by Khanna et al. [11]:

Lemma 11. $\alpha(G) \geq 2 - opt \geq \frac{1+\tau}{\Delta+2}n$.

For k-clique free graphs Halldórsson and Radhakrishnan [9] obtained improved bounds.

Lemma 12. *On a k-clique free graph G, $\alpha(G) \geq 2 - opt \geq \frac{2}{\Delta+k-1}n$.*

Remark: In [9] the weaker lower bound $\frac{2}{\Delta+k}n$ was shown. However, their proof admits the stronger lower bound stated above. Actually, if A is a maximal independent set, each vertex in $V - A$ has at least one neighbour in A. Since G is k-clique free, for each vertex $v \in A$, at most k-2 (instead of k-1) vertices can be adjacent only to u and some other vertices not in A.

The algorithm $CliqueRemoval_k$ then can be described as follows.

$CliqueRemoval_k$
$A_0 \leftarrow MIS(G)$
for $l = k$ downto 2 do
$S \leftarrow CliqueCollection(G, l)$
$G \leftarrow G - S$
$A_l \leftarrow l - clique - freeAlgorithm(G)$
od
Output A_i of maximum cardinality
end

Theorem 13. *$CliqueRemoval_p$, using 2-opt and l-clique-free achieves a performance ratio of at most*

$$[\frac{\Delta}{2} + 2 + \sum_{j=3}^{k} \frac{1}{j(j-1)f_j(\Delta)} + \frac{p}{2}(H_{p-1} - H_k + \frac{\Delta}{k})]/(p+1)$$

for graphs of maximum degree $\Delta \geq 3$ in polynomial time $O(n^k)$, where $p = \omega(G) + 1$.

Proof. Let n_t denote the number of vertices in the t-clique-free graph. Thus, $n \geq n_p \geq \ldots \geq n_3 \geq n_2 \geq 0$. From the approach of Nemhauser and Trotter [12] we may assume that $n_2 = 0$.

The size of the optimal solution is τn, which can be bounded by

$$\tau n \leq \frac{1}{2}(n_3 - n_2) + \ldots + \frac{1}{p}(n_{p+1} - n_p) = \sum_{i=3}^{p} \frac{1}{i(i-1)}n_i + \frac{1}{p}n. \qquad (17)$$

Then our algorithm is guaranteed to find an independent set of size at least

$$max[\frac{1+\tau}{\Delta+2}n, \max_{k+1 \leq t \leq p} \frac{2}{\Delta+t-1}n_t, f_3(\Delta)n_3, \ldots, f_k(\Delta)n_k]$$

by Lemma 11 and Lemma 12. Thus, the performance ratio ρ achieved by the algorithm is bounded by

$$\rho \le min[\frac{\tau n}{\frac{1+\tau}{\Delta+2}n}, \frac{\tau n}{\frac{2}{\Delta+t-1}n_t}, \frac{\tau n}{f_3(\Delta)n_3}, \dots, \frac{\tau n}{f_k(\Delta)n_k}].$$

As in [9] we derive from this , respectively, that

$$\tau \ge \frac{\rho}{\Delta+2-\rho}, \tag{18}$$

$$n_t \le \frac{\tau}{\rho} \cdot \frac{\Delta+t-1}{2}n, \ t=3,4,\dots,p \tag{19}$$

$$n_j \le \frac{\tau}{\rho} \cdot \frac{1}{f_j(\Delta)}n, \ j=3,\dots,k \tag{20}$$

Combining (17),(20) and (19), we find that

$$\tau \le \frac{\tau}{\rho}s_{\Delta,p} + \frac{1}{p}$$

where

$$s_{\Delta,p} = \sum_{j=3}^{k} \frac{1}{j(j-1)f_j(\Delta)} + \sum_{i=k+1}^{p} \frac{\Delta+i-1}{2i(i-1)}$$

$$= \sum_{j=3}^{k} \frac{1}{j(j-1)f_j(\Delta)} + \frac{1}{2}[(H_p - H_k) + \Delta(\frac{1}{k} - \frac{1}{p})].$$

Thus,

$$\tau \le \frac{1}{p(1 - s_{\Delta,p}/\rho)}. \tag{21}$$

Hence, from (18) and (21)

$$\frac{\rho}{\Delta+2-\rho} \le \frac{1}{p(1 - s_{\Delta,p}/\rho)}$$

which simplifies to

$$\rho \le \frac{\Delta+2+ps_{\Delta,p}}{p+1}$$

and we obtain the claimed inequality.

$\square$

If we now assume that Δ and p are growing functions, then the Δ/k term will dominate for a $\Delta/2k$ asymptotic ratio.

Corollary 14. *CliqueRemoval$_p$, using 2-opt and l-clique-free achieves a performance ratio of $\Delta/2k\ (1 + o(1))$ in polynomial time of $O(n^k)$.*

Acknowledgments. We thank the three anonymous referees for their helpful suggestions and corrections. Very recently M. M. Halldórsson kindly lead our attention to some related work in [15].

References

[1] M. Ajtai, P. Erdös, J. Komlós and E. Szemerédi, *On Turán's theorem for sparse graphs*, Combinatorica 1 (4) (1981) 313 - 317.

[2] S. Arora, C. Lund, R. Motwani, M. Sudan and M. Szegedy, *Proof verification and hardness of approximation problems*, Proc. 33rd IEEE FoCS, 1992, 14 - 23.

[3] P. Berman and M. Fürer, *Approximating maximum independent sets by excluding subgraphs*, Proc. Fifth ACM-SIAM Symp. on Discrete Algorithms, 1994, 365 - 371.

[4] R. B. Boppana and M. M. Halldórsson, *Approximating maximum independent set by excluding subgraphs*, BIT 32 (1992) 180 - 196.

[5] J. A. Bondy and U. S. R. Murty, *Graph Theory with Applications* (Macmillan, London and Elsevier, New York, 1976).

[6] Y. Caro, *New Results on the Independence Number*, Technical Report, Tel-Aviv University, 1979.

[7] M. R. Garey and D. S. Johnson, *Computers and Intractability, A Guide to the Theory of NP-Completeness*, W. H. Freeman and Company, New York, 1979.

[8] J. Håstad, *Clique is hard to approximate within $n^{1-\epsilon}$*, 37th Annual Symposium on Foundations of Computer Science, 1996, 627 - 636.

[9] M. M. Halldórsson and J. Radhakrishnan, *Improved approximations of Independent Sets in Bounded-Degree Graphs*, SWAT'94, LNCS 824 (1994) 195 - 206.

[10] M. M. Halldórsson and J. Radhakrishnan, *Greed is Good: Approximating Independent Sets in Sparse and Bounded-Degree Graphs*, Algorithmica 18 (1997) 145 - 163.

[11] S. Khanna, R. Motwani, M. Sudan and U. Vazirani, *On syntactic versus computational views of approximability*, Proc. 35th IEEE FoCS, 1994, 819 - 830.

[12] G. L. Nemhauser and L. E. Trotter. Jr., *Vertex Packings: Structural Properties and Algorithms*, Mathematical Programming 8 (1975) 232 - 248.

[13] J. B. Shearer, *A Note on the Independence Number of Triangle-Free Graphs*, Discrete Math. 46 (1983) 83 - 87.

[14] J. B. Shearer, *A Note on the Independence Number of Triangle-Free Graphs, II*, J. Combin. Ser. B 53 (1991) 300 - 307.

[15] J. B. Shearer, *On the Independence Number of Sparse Graphs*, Random Structures and Algorithms 5 (1995) 269 - 271.

[16] P. Turán, *On an extremal problem in graph theory* (in Hungarian), Mat. Fiz. Lapok 48 (1941) 436 - 452.

[17] V. K. Wei, *A Lower Bound on the Stability Number of a Simple Graph*, Technical memorandum, TM 81 - 11217 - 9, Bell laboratories, 1981.

Approximating an Interval Scheduling Problem

Frits C.R. Spieksma

Department of Mathematics, Maastricht University, P.O. Box 616, NL-6200 MD
Maastricht, The Netherlands, Tel.:+31-43-3883359, Fax:+31-43-3211889,
spieksma@math.unimaas.nl

Abstract. In this paper we consider a general interval scheduling problem. We show that, unless $\mathcal{P} = \mathcal{NP}$, this maximization problem cannot be approximated in polynomial time within arbitrarily good precision. On the other hand, we present a simple greedy algorithm that delivers a solution with a value of at least $\frac{1}{2}$ times the value of an optimal solution. Finally, we investigate the quality of an LP-relaxation of a formulation for the problem, by establishing an upper bound on the ratio between the value of the LP-relaxation and the value of an optimal solution.

1 Introduction

Consider the following problem. Given are n k-tuples of intervals on the real line, that is for each interval l a starting time s_l and a finishing time f_l $(> s_l)$ is known, $l = 1, \ldots, kn$. We assume that all starting and finishing times are integers. An interval is said to be *active* at time t iff $t \in [s_l, f_l)$. Two intervals *intersect* iff there is a time t during which both intervals are active. The problem is to select as many intervals as possible such that (i) no two selected intervals intersect, and (ii) at most one interval is selected from each k-tuple. A k-tuple of intervals is sometimes referred to as a *job*. We refer to this problem as the Job Interval Selection Problem with k intervals per job or JISPk for short. (Observe that an instance where the number of intervals per job is not the same for all jobs is easily transformed to an instance of JISPk for some k by duplicating intervals).

An alternative way of looking at JISPk is by adopting a graph-theoretical point of view. Indeed, let us construct a graph that has a node for each interval and in which two nodes are connected if the corresponding intervals belong to the same job (the *job* edges) or if the corresponding intervals intersect (the *intersection* edges). (Notice that an edge in this graph can be a job edge as well as an interval edge; this reflects the case when two intervals of a same job intersect). JISPk is now equivalent to finding a maximum stable set in this graph. Obviously, the graph induced by the job edges consists of n disjoint cliques of size k, and the graph induced by the intersection edges is an interval graph. Thus, the graph constructed is the edge union of an interval graph and a graph consisting of n disjoint cliques of size k. Notice that in case $k = 1$ the problem reduces to finding a maximum stable set in an interval graph.

JISPk belongs to the field of interval scheduling problems. These problems arise in a variety of settings. Here, we simply refer to Carter and Tovey (1992), Fischetti et al. (1992) and Kroon et al. (1997) and the references contained therein for examples of applications related to interval scheduling. JISPk is considered in Nakajima and Hakimi (1982) and in Keil (1992).

Keil (1992) proves that the problem of determing whether it is possible to select n intervals is $\mathcal{NP}$-complete for JISP3, whereas he shows that this question is solvable in polynomial time for JISP2. (This improved results in Nakajima and Hakimi (1982)). On the other hand, Kolen (1994) proved that given an integer K, the question whether one can select at least K intervals is already NP-complete for JISP2.

Our focus in this paper is on the following question: when restricting oneself to polynomial time algorithms, how good (in terms of quality of the solution) can one solve instances of JISPk, $k \geq 2$, in the worst case? Obviously, it follows from Keil (1992) that, unless $\mathcal{P} = \mathcal{NP}$, no polynomial time algorithm is able to solve JISPk exactly. Even more, we establish in Sect. 3 that, unless $\mathcal{P} = \mathcal{NP}$, no PTAS (see Sect. 2) exists for JISPk, for all $k \geq 2$. On the other hand we present in Sect. 4 a polynomial time approximation algorithm that delivers a solution with a value of at least $\frac{1}{2}$ times the value of an optimal solution. In Sect. 5 we formulate JISPk as an integer programming model and establish bounds on the value of the LP-relaxation in terms of the value of an optimal solution. For an overview of non-approximability results for 'classical' scheduling problems, we refer to Hoogeveen et al. (1997).

2 Preliminaries

A more extensive introduction to the issue of approximation and complexity can be found in Papadimitriou and Yannakakis (1991) and Crescenzi and Kann (1997). Here, we shortly list and describe some of the concepts we need.

- A *polynomial time ρ-approximation algorithm* for a maximization problem P is a polynomial time algorithm that, for all instances, outputs a solution with a value that is at least equal to ρ times the value of an optimal solution of P.
- A *polynomial time approximation scheme* (PTAS) is a family of polynomial time $(1 - \epsilon)$-approximation algorithms for all $\epsilon > 0$.
- An L-reduction. Given two maximization problems A and B, an L-reduction from A to B is a pair of functions R and S such that:
  - R and S are computable in polynomial time,
  - For any instance I of A with optimum cost $OPT(I)$, $R(I)$ is an instance of B with optimum cost $OPT(R(I))$, such that

$$OPT(R(I)) \leq \alpha \cdot OPT(I), \tag{1}$$

  for some positive constant α.

- For any feasible solution s of $R(I)$, $S(s)$ is a feasible solution of I such that

$$OPT(I) - c(S(s)) \leq \beta \cdot (OPT(R(I)) - c(s)), \qquad (2)$$

for some positive constant β, where $c(S(s))$ and $c(s)$ denote the costs of $S(s)$ and s respectively.

An L-reduction is an *approximation preserving* reduction, that is, if problem B can be approximated within $1 - \epsilon$ then problem A can be approximated within $1 - \alpha\beta\epsilon$ (assuming that there is an L-reduction from A to B).

- The class MAX SNP is a class that contains optimization problems that are approximable in polynomial time within a constant factor.
- The problem *Maximum Bounded 3-Satisfiability* (MAX-3SAT-B):

Input: A set of Boolean variables $X = \{x_1, x_2, \ldots, x_n\}$ and a set $C = \{C_1, C_2, \ldots, C_r\}$ of clauses over X. Each clause C_j ($j = 1, \ldots, r$) consists of precisely three literals and each variable x_i ($i = 1, \ldots, n$) occurs at most three times in C (either as literal x_i or as literal $\bar{x}_i$).

Goal: Find a truth assignment for the variables such that the number of satisfied clauses in C is maximum.

Measure: The number of satisfied clauses in C.

Papadimitriou and Yannakakis (1991) proved the following result:

Lemma 2.1. *MAX-3SAT-B is MAX SNP-hard.*

Arora et al. (1992) proved the following result:

Lemma 2.2. *If there exists a PTAS for some MAX SNP-hard problem, then $\mathcal{P} = \mathcal{NP}$.*

We now have sketched the tools that enable us to prove that JISPk has no PTAS (unless $\mathcal{P} = \mathcal{NP}$): this can be done by exhibiting an L-reduction from MAX-3SAT-B and using Lemma's 2.1 and 2.2.

3 A Non-approximability Result

Theorem 3.1. *JISPk does not have a PTAS unless $\mathcal{P} = \mathcal{NP}$ for each fixed $k \geq 2$.*

Proof. We prove the theorem by presenting an L-reduction from MAX-3SAT-B to JISP2. The result then follows from Lemma 2.1 and Lemma 2.2. Recall that $C = \{C_1, C_2, \ldots, C_r\}$ is a set consisting of r disjunctive clauses, each containing exactly 3 literals. Let $x_1, x_2, \ldots, x_n$ denote the variables in the r clauses and, for each $i = 1, \ldots, n$, let $m(i)$ denote the number of occurrences of variable x_i (either as literal x_i or as literal $\bar{x}_i$). Arbitrarily index the occurrences of variable x_i as occurrence $1, 2, \ldots, m(i)$. Notice that without loss of generality we can assume that each variable occurs at least twice in C, thus we have $2 \leq m(i) \leq 3$ for all i and that $\sum_i m(i) = 3r$.

We now construct an instance of JISP2, that is a graph $G = (V, E)$ which is the edge union of an interval graph and a matching. Let I denote an instance of MAX-3SAT-B and $R(I)$ the corresponding instance of JISP2 with corresponding optimal values $OPT(I)$ and $OPT(R(I))$. For each variable x_i in I, $i = 1, \ldots, n$, we have a subgraph $H1_i = (V1_i, E1_i)$ in $R(I)$, where $V1_i = \{T_{i1}, F_{i1}, T_{i2}, F_{i2}, \ldots, T_{i,m(i)}, F_{i,m(i)}\}$ and $E1_i = \{\{T_{ij}, F_{ij}\} \cup \{T_{ij}, F_{i,j+1}\}|\ j = 1, \ldots, m(i)\}$ (indices modulo $m(i)$). So for each variable x_i in I we have a cycle consisting of $2m(i)$ nodes in $R(I)$ (see Fig. 1).

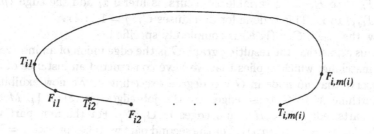

Fig. 1. The subgraph $H1_i$.

When no ambiguity is likely to arise, we refer to the nodes T_{ij} (F_{ij}), $j = 1, \ldots, m(i)$, in subgraph $H1_i$ as T-nodes (F-nodes).

For each clause C_j in I, $j = 1, \ldots, r$, we have a subgraph $H2_j = (V2_j, E2_j)$ in $R(I)$ as depicted in Fig. 2.

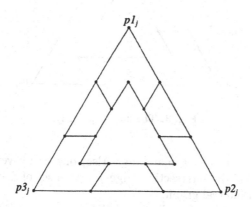

Fig. 2. The subgraph $H2_j$.

Again, when no ambiguity is likely to arise, we refer to the nodes $p1_j, p2_j$ and $p3_j$ in subgraph $H2_j$ as p-nodes. Notice that the size of a maximum stable set in the graph $H2_j$ is bounded by 8; moreover, if one is not allowed to use

p-nodes in a stable set, no more than 7 nodes from $H2_j$ can be in a stable set, $j = 1, \ldots, r$.

To connect the subgraphs introduced sofar in $R(I)$, consider some clause C_j, and consider the first variable occurring in this clause C_j, say x_i. Let this be the q-th occurrence of this variable x_i in C, $q \in \{1, 2, 3\}$. If the variable x_i occurs as literal x_i add the edge $\{p1_j, F_{iq}\}$ to E. If the variable x_i occurs as literal $\bar{x}_i$ add the edge $\{p1_j, T_{iq}\}$ to E. Consider now the second (third) variable occurring in C_j, say x_l, and let this be the q-th occurrence of this variable x_l in C, $q \in \{1, 2, 3\}$. If the variable x_l occurs as literal x_l add the edge $\{p2_j, F_{lq}\}$ ($\{p3_j, F_{lq}\}$) to E. If the variable x_l occurs as literal $\bar{x}_l$ add the edge $\{p2_j, T_{lq}\}$ ($\{p3_j, T_{lq}\}$) to E. This is done for all clauses C_j, $j = 1, \ldots, r$.

Now the graph $G = (V, E)$ is completely specified.

Let us argue that the resulting graph G is the edge union of an interval graph and a matching, which implies that we have constructed an instance of JISP2.

Observe that no node in G has degree exceeding 3. We now exhibit a perfect matching M in G; these edges are the job edges (see Sect. 1). M consists of two parts: edges in $\cup_i H1_i$ and edges in $\cup_j H2_j$. For the first part we take $\cup_i \{\{T_{ij}, F_{ij}\} \mid j = 1, \ldots, m(i)\}$. For the second part we take, for each $j = 1, \ldots, r$, the bold edges depicted in Fig. 3.

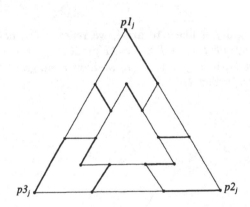

Fig. 3. The subgraph $H2_j$.

Obviously, M is indeed a matching. Also, one easily verifies that the remaining edges in G (the intersection edges) form a set of disjoint paths, which corresponds to an interval graph.

In order to show that this reduction fulfills inequalities (1) and (2), consider the following. Observe that $v \equiv OPT(I) \geq \frac{1}{2} r$. (Indeed, by considering the assignment: all variables true, and: all variables false, it follows that each clause is true in at least in one of both assignments). We have:

$$OPT(R(I)) \leq 3r + 8r = 11r \leq 22v = 22 \cdot OPT(I),$$

which proves (1). The first inequality follows from the fact that

- at most $m(i)$ nodes can be selected from each $H1_i$, $i = 1, \ldots, n$ (see Fig. 1), and $\sum_i m(i) = 3r$, and
- at most 8 nodes can be selected from each subgraph $H2_j$, $j = 1, \ldots, r$ (see Fig. 2).

To establish (2), we do the following. Consider an arbitrary solution to $R(I)$, that is any stable set s in G with size $c(s)$. We will map this solution s using intermediate solutions s' and s'' to a solution of MAX-3SAT-B, called $S(s)$. To do this we need the following definition. A stable set s in G is called *consistent* iff for each $i = 1, \ldots, n$, $m(i)$ nodes from $V1_i$ are in s.

Now we state a procedure which takes as input a stable set s. The output of the procedure is a consistent stable set called s' with the property that $c(s') \geq c(s)$.

Procedure

Consider s. For $i = 1, \ldots, n$, consider $V1_i$. There are two possibilities.

1. $m(i)$ nodes from $V1_i$ are in s. Then either all T-nodes or all F-nodes from $V1_i$ are in s and we leave s unaltered.
2. Less than $m(i)$ nodes from $V1_i$ are in s. Let cT (cF) be the number of T-nodes (F-nodes) in $V1_i$ that are connected to p-nodes that are in s. (Notice that $cT + cF \leq 3$.) Distinguish two subcases:
   - If $cT > cF$ ($cT < cF$), it follows that $cF \leq 1$ ($cT \leq 1$). Modify s by selecting all F-nodes from $V1_i$ (and undo the selection of any T-nodes in s), and, if $cF = 1$, undo in s the selection of the p-node connected to an F-node. Notice that this modification does not decrease the number of nodes in the stable set.
   - $cT = cF$. In that case, select from $V1_i$ all T-nodes, and undo the selection of a p-node connected to a T-node. (Notice that there can be at most 1 such node). Again, modifying s in this way does not decrease the number of selected nodes.

End of Procedure

After applying this procedure to any stable set s in G, a consistent solution s' is delivered. Now we describe how to modify s' to get solution s''. Consider in s' those subgraphs $H2_j$ whose corresponding p-nodes all three cannot be chosen, due to nodes from $\cup_i V1_i$ in s'. Suppose there are $r - l$ of those subgraphs in s'. Then we modify s' such that in l subgraphs $H2_j$ 8 nodes are selected and in $r - l$ subgraphs $H2_j$ 7 nodes (this is always possible, see Figs. 1 and 2). This gives us a consistent solution s'' with $c(s'') \geq c(s')$. Since s'' is consistent, it is now straightforward to identify the corresponding solution $S(s)$ in MAX-3SAT-B: simply set variable x_i, $i = 1, \ldots, n$ true if all T-nodes in subgraph $H1_i$ are selected in s'', else set x_i false. How many clauses in I are satisfied by this truth assignment? Observe that the construction of G implies that if for some consistent stable set s each p-node from some $H2_j$ is connected to a node in

$\cup_i V1_i$ that is in s, then the corresponding truth assignment renders clause C_j not satisfied, and vice versa. Thus, by the construction of s'', it follows that a subgraph $H2_j$ for which 7 nodes are in s'' corresponds to a not satisfied clause, and otherwise the clause is satisfied, $j = 1, \ldots, r$. This implies that l clauses in I are satisfied by this truth assignment.

Again, let $v = OPT(I)$, and let $c(S(s)) = l$. The following (in)equalities are true:

- $c(s) \leq c(s'')$ (by construction),
- $c(s'') = 3r + 8l + 7(r - l) = 10r + l$ (by construction), and
- $OPT(R(I)) \geq 3r + 8v + 7(r - v) = 10r + v$ (consider the truth assignment that is optimum for I; evidently, we can exhibit in $R(I)$) a corresponding stable set of size $10r + v$).

Thus

$$OPT(R(I)) - c(s) \geq OPT(R(I)) - c(s'') \geq$$
$$10r + v - (10r + l) = v - l = OPT(I) - c(S(s)),$$

which proves (2). □

This reduction is based on an NP-completeness proof in Kolen (1994) (which in turn was inspired by a reduction in Garey et al. (1976)). There are a number of implications that can be observed from this reduction. First of all, the reduction remains valid if there are restrictions on the number of intervals that is active at time t for some t. More specifically, let $\omega_t(I)$ be the number of intervals in I that is active at time t, and define the *maximum intersection* as $\omega(I) = \max_t \omega_t(I)$. Notice that Theorem 3.1 remains true even when $\omega(I) \leq 2$ (whereas the problem becomes trivial when $\omega(I) \leq 1$). Also, the reduction remains valid for short processing times. Indeed, even if $f_l - s_l = 2$ for all intervals l, Theorem 3.1 remains true (whereas the problem again becomes trivial in the case that $f_l - s_l = 1$ for all l). Finally, observe the following. As mentioned in Sect. 1, Keil (1992) proves that the question whether one can select n intervals in a JISP2 instance is solvable in polynomial time. In fact, this result can also be proved as follows. Graphs for which the size of a maximum matching equals the size of a minimum vertex cover are said to have the *König property*. Since the complement of a minimum vertex cover is a maximum stable set, it follows that for graphs with the *König property* the cardinality of a maximum stable set can be found in polynomial time. Now, the size of a maximum matching for a graph corresponding to a JISP2 instance equals n. So the question Keil (1992) answered is equivalent to the question whether the graph corresponding to a JISP2 instance has the König property. This problem can be solved in polynomial time (see Plummer (1993) and the references contained therein).

4 An Approximation Algorithm

In this section we describe a simple 'from-left-to-right' algorithm, and show that it is $\frac{1}{2}$-approximation algorithm. The algorithm can informally be described as follows: start "at the left", and take, repeatedly, the earliest ending feasible interval. Applied to any instance I of JISPk, this gives at least $\frac{1}{2}OPT(I)$. A more formal description of the algorithm, referred to as GREEDY, is as follows. Let $G(I)$ be the set of intervals selected from I by GREEDY, and let $J(i)$ be the set of intervals belonging to the job corresponding to interval i, $i = 1, \ldots, kn$.

GREEDY:
$T := -\infty$;
$G(I) := \emptyset$;
$S :=$ set of all intervals in I;
while $\max_{i \in S} s_i \geq T$ do
begin

$\qquad i^* := \arg(\min_{i \in S}\{f_i|\ s_i \geq T\})$ (break ties arbitrarily);
$\qquad G(I) := G(I) \cup \{i^*\}$;
$\qquad S := S \setminus J(i^*)$;
$\qquad T := f_{i^*}$;
end;

Obviously, GREEDY is a polynomial time algorithm.

Theorem 4.1. GREEDY is a $\frac{1}{2}$-approximation algorithm for JISPk, $k \geq 1$. Moreover, there exist instances of JISPk for which this bound is tight, for all $k \geq 2$.

Proof. Consider some instance I of JISPk. Applying GREEDY gives us a solution with $|G(I)|$ intervals selected. The idea of the proof is to partition I into two instances I_1 and I_2, and show that for each of those instances it is impossible to select more than $|G(I)|$ intervals. Clearly, then no more than $2|G(I)|$ intervals can be selected, proving the first part of the theorem.

Now, let I_1 consist of the jobs whose intervals are selected by GREEDY, and let I_2 consist of all other jobs. Obviously, $OPT(I_1) \leq |G(I)|$, since I_1 contains no more as $|G(I)|$ jobs. Let the finishing times of all intervals selected by GREEDY be indexed $e_1 < e_2 < \ldots < e_{|G(I)|}$ and let $e_0 = -\infty$. For each interval in I_2 we know that it is active at $e_j - 1$ for some $j = 1, \ldots, |G(I)|$. (Otherwise it would have been selected by GREEDY). In other words, all intervals in I_2 that have a starting time in $[e_{j-1}, e_j)$ have a finishing time after time e_j, $j = 1, \ldots, |G(I)|$. Thus at most one of those can be in a solution of I_2. Since there are only $|G(I)|$ such time-intervals $[e_{j-1}, e_j)$, at most $|G(I)|$ intervals can be selected. Summarizing, we have: $OPT(I) \leq OPT(I_1) + OPT(I_2) \leq |G(I)| + |G(I)|$.

To show that this is best possible for GREEDY, consider the instance of JISP2 depicted in Fig. 4 (where the interval corresponding to job 2 has multiplicity 2).
It is easy to see that for this instance I, $OPT(I) = 2$, whereas $|G(I)| = 1$. □

Remark 4.1. Notice that, for $k = 1$, GREEDY reduces to a special case of an algorithm described by Carlisle and Lloyd (1995) and Faigle and Nawijn (1995), and hence always finds an optimal solution.

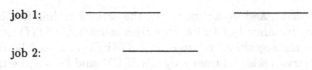

job 1:

job 2:

Fig. 4. A worst-case instance for GREEDY.

5 An IP-formulation for JISPk

Consider now the following Integer Programming formulation (IP) for JISPk which assumes wlog that job i consists of intervals $k(i-1)+1, k(i-1)+2, \ldots, ki$. Let $x_l = 1$ if interval l is selected and 0 otherwise, and let $A(l) = \{j : \text{interval } j \text{ is active at } f_l - 1\}$, $l = 1, \ldots, kn$. (Notice that $l \in A(l)$).

(IP) Maximize $\sum_{l=1}^{kn} x_l$

$$\text{subject to } \quad x_{k(i-1)+1} + \ldots + x_{ki} \leq 1 \quad \text{for all } i = 1, \ldots, n, \qquad (3)$$

$$\sum_{j \in A(l)} x_j \leq 1 \qquad \text{for all } l = 1, \ldots, kn, \qquad (4)$$

$$x_l \in \{0, 1\} \qquad \text{for all } l = 1, \ldots, kn. \qquad (5)$$

Constraints (3) express that at most 1 interval per job can be selected, while constraints (4) ensure that no intersection occurs in the set of selected intervals. Constraints (5) are the integrality constraints. Let $v_{LP}(I)$ denote the value of the LP-relaxation of (IP) with respect to instance I of JISPk.

Theorem 5.1. $v_{LP}(I) \leq 2 \cdot OPT(I)$ *for all* I. *Moreover, this bound is asymptotically tight.*

Proof. The idea is as follows. Let us construct a solution which is feasible to the dual of the LP-relaxation of (IP). This solution will have a value, say $v_D(I)$, bounded by $2 \cdot OPT(I)$ for all I. Then, by LP-duality we are done: $v_{LP}(I) \leq v_D(I) \leq 2 \cdot OPT(I)$ for all I.

Associating z-variables to the first set of constraints of (IP) and y-variables to the second set of constraints, we get the following dual of the LP-relaxation of (IP) (let $A^{-1}(l) = \{j : \text{interval } l \text{ is active at } f_j - 1\}, l = 1, \ldots, kn$):

(D) Minimize $\sum_{l=1}^{kn} y_l + \sum_{i=1}^{n} z_i$

$$\text{subject to } z_{\lceil l/k \rceil} + \sum_{j \in A^{-1}(l)} y_j \geq 1 \quad \text{for all } l = 1, \ldots, kn,$$

$$\text{all variables } \geq 0.$$

One can think of the z-variables as horizontal lines, such that z_i 'touches' intervals $k(i-1)+1, \ldots, ki$, $(i = 1, \ldots, n)$ and of the y-variables as vertical lines such that y_l is at time $f_l - 1$ and 'touches' all intervals in $A(l)$ $(l = 1, \ldots, kn)$. The dual problem (D) is now to give the dual variables nonnegative weights such that total weight is minimized and every interval receives at least weight 1 from those dual variables by which it is touched.

Consider now the set of all optimal solutions to (IP) with respect to some instance I (so the optimal integral solutions), and consider for each optimal solution the increasing sequence of finishing times of intervals selected in that optimal solution. Let $EARLYOPT(I)$ be the set of intervals in I corresponding to finishing times in the lexicographic smallest sequence. Let $SISTERS(I)$ be the set of intervals whose corresponding jobs have an interval in $EARLYOPT(I)$ (formally $SISTERS(I) = \cup_i(J(i) \setminus i)$), and let $REST(I)$ be the set of all remaining intervals. Thus, we have partitioned the set of intervals in I into three subsets.

We now construct the following dual solution:

1) $y_l = 1$ for all $l \in EARLYOPT(I)$. Notice that the construction leading to $EARLYOPT(I)$ implies that each interval from $REST(I)$ is touched by some y_l, $l \in EARLYOPT(I)$. (Indeed, suppose not, then there exists an "earlier" optimal solution than $EARLYOPT(I)$ which is impossible.) Thus, by choosing these weights, each interval from $EARLYOPT(I)$ as well as each interval from $REST(I)$ receives weight 1. Total weight spent: $OPT(I)$.

2) $z_{\lceil \frac{l}{k} \rceil} = 1$ for all $l \in EARLYOPT(I)$. This implies that each interval from $EARLYOPT(I)$ as well as from $SISTERS(I)$ receives weight 1. Total weight spent: $OPT(I)$.

3) All other dual variables are 0.

It is easy to verify that this constitutes a feasible dual solution with weight $2 \cdot OPT(I)$. The first part of the theorem then follows. To establish the second part, consider the following instance, depicted in Fig. 5 (where the interval corresponding to job 2 has multiplicity k).

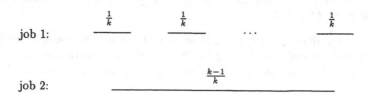

Fig. 5. An instance of JISPk.

It is not hard to verify that the numbers above the intervals in Fig. 5 are the optimal LP-values of the corresponding x-variables. Thus, for this instance we have that $v_{LP}(I) = 2 - \frac{1}{k}$, whereas $OPT(I)$ clearly equals 1. $\qquad\square$

Remark 5.1. Notice that we actually proved a slightly stronger statement than announced in Theorem 5.1. Indeed, let $v_{DIP}(I)$ be the value corresponding to the formulation which arises when to problem D the constraints $y, z \in \{0, 1\}$ are added. Arguments in the proof of Theorem 5.1 imply that $v_{DIP}(I) \leq 2 \cdot OPT(I)$ and the instance in Fig. 5 shows that this inequality is tight for each $k \geq 2$.

Although the bound in Theorem 5.1 is asymptotically tight, there remains a sizable gap for JISPk instances with small values of k (for $k = 2$, the gap is $\frac{3}{2}$ versus 2). The following theorem closes part of this gap.

Theorem 5.2. $v_{LP}(I) \leq \frac{5}{3} \cdot OPT(I)$ for all JISP2 instances I.

Proof. We refine the proof of Theorem 5.1. Construct the following dual solution.

1) $y_l = \frac{2}{3}$ for all $l \in EARLYOPT(I)$. It follows (see the proof of Theorem 5.1) that each interval from $EARLYOPT(I)$ as well as each interval from $REST(I)$ receives weight $\frac{2}{3}$. Total weight spent: $\frac{2}{3}OPT(I)$.
2) $z_{\lceil \frac{l}{2} \rceil} = \frac{1}{3}$ for all $l \in EARLYOPT(I)$. This implies that each interval from $EARLYOPT(I)$ as well as from $SISTERS(I)$ receives weight $\frac{1}{3}$. Total weight spent: $\frac{1}{3}OPT(I)$.

To proceed, we construct from instance I an instance I' by deleting from I all intervals in $EARLYOPT(I)$. Obviously, $OPT(I') \leq OPT(I)$.

3) $y_l = \frac{1}{3}$ for all $l \in EARLYOPT(I')$. Notice that each interval from $EARLYOPT(I')$ as well as each interval from $REST(I')$ receives weight $\frac{1}{3}$. Total weight spent: at most $\frac{1}{3}OPT(I)$.

Construct now the instance I'' by taking all intervals from $SISTERS(I)$ and $SISTERS(I')$. Observe that there are no 2 intervals present in I'' belonging to a same job. Thus, we are now dealing with finding a maximum stable set on an interval graph. Such an instance is solvable by **GREEDY** as explained earlier. Set

4) $y_l = \frac{1}{3}$ for all $l \in G(I'')$. Notice that each interval from I'' is touched by some y_l, $l \in G(I'')$. Notice also that $|G(I'')| \leq OPT(I)$, thus total weight spent: at most $\frac{1}{3}OPT(I)$.

All other dual variables get weight 0. If we sum total weight spent in 1)-4) it follows we have spent not more as $\frac{5}{3}$ OPT. It remains to argue that each interval from the instance has received weight at least 1. Take any interval from I and distinguish 5 cases:

i: It belongs to $EARLYOPT(I)$. Then it gets $\frac{2}{3}$ from 1) and $\frac{1}{3}$ from 2).
ii: It belongs to $SISTERS(I)$. Then it gets $\frac{1}{3}$ from 2), $\frac{1}{3}$ from 3) (since each interval from $SISTERS(I)$ is either an $EARLYOPT(I')$ or a $REST(I')$ interval) and it gets $\frac{1}{3}$ from 4).
iii: It belongs to $REST(I)$ and $EARLYOPT(I')$. Then it gets $\frac{2}{3}$ from 1) and $\frac{1}{3}$ from 3).
iv: It belongs to $REST(I)$ and $REST(I')$. Then it gets $\frac{2}{3}$ from 1) and $\frac{1}{3}$ from 3).
v: It belongs to $REST(I)$ and $SISTERS(I')$. Then it gets $\frac{2}{3}$ from 1) and $\frac{1}{3}$ from 4).

This completes the proof. $\qquad \square$

References

Arora, S., Lund, C., Motwani, R., Sudan, M., Szegedy, M.: Proof verification and hardness of approximation problems. Proceedings of the 33rd IEEE Symposium on the Foundations of Computer Science (1992) 14–23

Carlisle, M.C., Lloyd, E.L.: On the k-coloring of intervals. Discrete Applied Mathematics **59** (1995) 225–235

Carter, M.W., Tovey, C.A.: When is the classroom assignment problem hard? Operations Research **40** (1992) S28–S39

Crescenzi, P., Kann, V.: A compendium of NP optimization problems. http://www.nada.kth.se/nada/~viggo/problemlist/compendium.html

Faigle, U., Nawijn, W.M.: Note on scheduling intervals on-line. Discrete Applied Mathematics **58** (1995) 13–17

Fischetti, M., Martello, S., Toth, P.: Approximation algorithms for fixed job schedule problems. Operations Research **40** (1992) S96–S108

Garey, M.R., Johnson, D.S., Stockmeyer, L.: Some simplified NP-complete graph problems. Theoretical Computer Science **1** (1976) 237–267

Hoogeveen, J.A., Schuurman, P., Woeginger, G.J.: Non-approximability results for scheduling problems with minsum criteria. Eindhoven University of Technology, COSOR Memorandum 97-24, to appear in the Proceedings of the 6th IPCO Conference, Houston.

Keil, J.M.: On the complexity of scheduling tasks with discrete starting times. Operations Research Letters **12** (1992) 293–295

Kolen, A.W.J., personal communication.

Kroon, L.G., Salomon, M., van Wassenhove, L.N.: Exact and approximation algorithms for the tactical fixed interval scheduling problem. Operations Research **45** (1997) 624–638

Nakajima, K., Hakimi, S.L.: Complexity results for scheduling tasks with discrete starting times. Journal of Algorithms **3** (1982) 344–361

Papadimitriou, C.H., Yannakakis, M.: Optimization, approximation and complexity classes. Journal of Computer and System Sciences **43** (1991) 425–440

Plummer, M.D.: Matching and vertex packing: how "hard" are they? Annals of Discrete Mathematics **55** (1993) 275–312

Finding Dense Subgraphs with Semidefinite Programming

Anand Srivastav[1] and Katja Wolf[2]

[1] Mathematisches Seminar, Christian-Albrechts-Universität zu Kiel,
Ludewig-Meyn-Str. 4, D-24098 Kiel, Germany
(asr@numerik.uni-kiel.de)
[2] Zentrum für Paralleles Rechnen, Universität zu Köln,
Weyertal 80, D-50931 Köln, Germany
(wolf@zpr.uni-koeln.de)

Extended Abstract

Abstract. In this paper we consider the problem of computing the heaviest k-vertex induced subgraph of a given graph with nonnegative edge weights. This problem is known to be $\mathcal{NP}$-hard, but its approximation complexity is not known. For the general problem only an approximation ratio of $\tilde{\mathcal{O}}(n^{0.3885})$ has been proved (Kortsarz and Peleg (1993)). In the last years several authors analyzed the case $k = \Omega(n)$. In this case Asahiro et al. (1996) showed a constant factor approximation, and for dense graphs Arora et al. (1995) obtained even a polynomial-time approximation scheme. We give a new approximation algorithm for arbitrary graphs and $k = n/c$ for $c > 1$ based on semidefinite programming and randomized rounding which achieves for some c the presently best (randomized) approximation factors.

Key Words. Subgraph Problem, Approximation Algorithms, Randomized Algorithms, Semidefinite Programming.

1 Introduction

For an undirected graph $G = (V, E)$ with nonnegative edge weights w_{ij} for $(i, j) \in E$ and an integer $k \leq n = |V|$ the HEAVIEST SUBGRAPH problem is to determine a subset S of k vertices such that the weight of the subgraph induced by S is maximized. We measure the weight of the subgraph by computing $\omega(S) = \sum_{i \in S, j \in S} w_{ij}$. (For convenience, we set $w_{ij} = 0$ for $(i, j) \notin E$, implicitly assuming that G is a complete graph.) The unweighted case of the problem ($w_{ij} = 1$ for $(i, j) \in E$) is called DENSEST SUBGRAPH. These problems arise in several applications. (See [4, 15] for a detailed discussion.)

Both problems are $\mathcal{NP}$-hard, which can be easily seen by a reduction from MAXIMUM CLIQUE. The HEAVIEST SUBGRAPH problem remains $\mathcal{NP}$-hard when the weights satisfy the triangle inequality [15]. A promising and often successful approach to cope with the hardness of a combinatorial optimization problem is to design polynomial-time approximation algorithms.

Given an instance I of a maximization problem and an (approximation) algorithm A the *approximation ratio* $R_A(I)$ is defined by $R_A(I) = OPT(I)/A(I) \geq 1$, while $r_A(I) = 1/R_A(I) \leq 1$ is called the *approximation factor*. We will use both notations, but for the comparison of constant-factor approximations r_A will be more convenient.

Previous Work. For the case where the weights satisfy the triangle inequality Hassin, Rubinstein and Tamir [12] describe an algorithm which is similar to a greedy solution for constructing a maximum matching by repeatedly choosing the heaviest edge. Their algorithm has approximation ratio 2.

Arora, Karger and Karpinski [3] model the DENSEST SUBGRAPH problem as a quadratic 0/1 program and apply random sampling and randomized rounding techniques resulting in a polynomial-time approximation scheme (that is, a family of algorithms A_ε with approximation ratio $(1 + \varepsilon)$ for each $\varepsilon > 0$) for problem instances satisfying $k = \Omega(n)$ and $|E| = \Omega(n^2)$, or for instances where each vertex has degree $\Omega(n)$.

The general weighted problem without triangle inequality restrictions is considered in [2, 9, 14]. Kortsarz and Peleg [14] devise an approximation algorithm which achieves an approximation ratio of $\tilde{\mathcal{O}}(n^{0.3885})$. Asahiro et al. [2] analyze a greedy heuristic which repeatedly deletes a vertex with the least weighted degree from the current graph until k vertices are left. They derive the following bounds for the approximation ratio R_{greedy}

$$\left(\tfrac{1}{2} + \tfrac{n}{2k}\right)^2 - \mathcal{O}(1/n) \leq R_{\text{greedy}} \leq \left(\tfrac{1}{2} + \tfrac{n}{2k}\right)^2 + \mathcal{O}(1/n) \quad \text{for} \quad n/3 \leq k \leq n,$$
$$2\left(\tfrac{n}{k} - 1\right) - \mathcal{O}(1/k) \leq R_{\text{greedy}} \leq 2\left(\tfrac{n}{k} - 1\right) - \mathcal{O}(n/k^2) \quad \text{for} \quad k < n/3.$$

Goemans [9] studies a linear relaxation of the problem. Linear programming yields a fractional solution, subsequent randomized rounding gives an integer solution which may exceed the allowed number of vertices but this is repaired in a greedy manner. His algorithm has expected approximation ratio $2 + \mathcal{O}(n^{-1/2})$ for $k = n/2$.

Independently of our work, Feige and Seltser [7] have developed an algorithm which is based on a different semidefinite programming relaxation and uses a norm-based rounding procedure while ours takes directions of vectors into account. Their approximation ratio is roughly n/k. For $k \simeq n^{1/3}$ they point out the limits of the method in comparison to [14].

The Results. We present a randomized rounding algorithm for arbitrary graphs and $k = n/c, c > 1$, which outputs for every sufficiently small $\varepsilon > 0$ a k-vertex subgraph of expected weight at least $r(c)HS_{\text{opt}}$ where HS_{opt} is the value of an optimal solution and

$$r(c) = (1 - c/n)(1 - \varepsilon)^2 (\beta/c^2 + \frac{(c^2 - 1)(c - 1)(1 - \varepsilon)\alpha\beta}{c^4(c - \alpha + \varepsilon) - c^2(c^2 - 1)(c - 1)(1 - \varepsilon)\alpha})$$

($\alpha > 0.87856$ and $\beta > 0.79607$ are the constants derived in the approximation algorithms by Goemans and Williamson for MAXCUT and MAXDICUT [10].)

The following table shows the values of r for some interesting values of k (assuming n to be sufficiently large and ε sufficiently small so that the factor $(1-c/n)(1-\varepsilon)^2$ is negligible). We have listed the expected approximation factor for choosing a random subgraph in the first column, the approximation factor $r_{\text{greedy}} = 1/R_{\text{greedy}}$ in the second column and our approximation factor $r(c)$ in the third column.

k	random	r_{greedy}	$r(c)$
$n/2$	0.25	$0.\overline{4}$	0.4825
$n/3$	$0.\overline{1}$	0.25	0.3353
$n/4$	0.0625	$0.1\overline{6}$	0.2387

Note that in all cases shown in the table we have an improvement on the approximation factors due to Asahiro et al. [2]. For $k = n/2$ our factor is slightly smaller than the factor of 0.5 achieved by Goemans [9] and Feige and Seltser [7], while for $k = n/3$ our factor is better. An example due to M. Langberg for $k = n/2$ shows that in this case the approximation guarantee of our relaxation cannot be better than 0.5.

The paper is organized as follows. In Section 2 we show how the HEAVIEST SUBGRAPH problem can be formulated as a quadratic program. A relaxation of the program can be solved within any desired precision in polynomial time with the help of semidefinite programming. Section 3 is dedicated to the analysis of the expected approximation factor of the algorithm for $k = n/c, c > 1$. We conclude in Section 4 and outline how the approximation factor could be further improved.

2   Modeling the Problem as a Semidefinite Program

In this section we will derive a suitable formulation for the HEAVIEST SUBGRAPH as a nonlinear program and state our approximation algorithm. We introduce a variable x_i for each vertex $i \in V = \{1, \ldots, n\}$ and, in addition, another variable x_0 to express whether a vertex belongs to the subgraph or not.

$$i \in S \Leftrightarrow x_0 x_i = 1.$$

The optimal value for the HEAVIEST SUBGRAPH can be obtained as a solution to the following program

$$\text{maximize } \tfrac{1}{4} \sum_{(i,j) \in E} w_{ij} (1 + x_0 x_i)(1 + x_0 x_j)$$

$$\text{subject to } \sum_{i=1}^{n} x_0 x_i = 2k - n \qquad (HS)$$

$$x_0, x_1, \ldots, x_n \in \{-1, 1\}.$$

The term $(1 + x_0 x_i)(1 + x_0 x_j)/4 = (1 + x_0 x_i + x_0 x_j + x_i x_j)/4$ evaluates to 1 if i and $j \in S$ and to 0 otherwise. Thus, an edge (i, j) having both endpoints in the induced subgraph on S contributes w_{ij} to the objective function. The constraints guarantee that this subgraph has size k. Since it is $\mathcal{NP}$-hard to find a solution to this integer program, we relax the integrality constraint and permit the variables to be vectors in the unit sphere in $\mathbb{R}^{n+1}$. The product is replaced by the inner product of two vectors. Let B_1 be the unit sphere in $\mathbb{R}^{n+1}$

$$B_1 = \{x \in \mathbb{R}^{n+1} \mid \| x \|_2 = 1\}.$$

$$\text{maximize } \frac{1}{4} \sum_{(i,j) \in E} w_{ij} (1 + x_0 \cdot x_i + x_0 \cdot x_j + x_i \cdot x_j)$$

$$\text{subject to } \quad \sum_{i=1}^{n} x_0 \cdot x_i = 2k - n \qquad (SDP)$$

$$x_0, x_1, \ldots, x_n \in B_1.$$

Using the variable transformation $y_{ij} := x_i \cdot x_j$ we may translate the above program into an equivalent semidefinite program.

$$\text{maximize} \quad \frac{1}{4} \sum_{(i,j) \in E} w_{ij} (1 + y_{0i} + y_{0j} + y_{ij})$$

$$\text{subject to} \quad \sum_{i=1}^{n} y_{0i} = 2k - n$$

$$y_{ii} = 1 \text{ for } i = 0, \ldots, n$$

$$Y = (y_{ij}) \text{ symmetric and positive semidefinite.}$$

This program can be solved within an additive error of δ of the optimum in time polynomial in the size of the input and $\log(1/\delta)$ by, for example, interior-point algorithms or the ellipsoid method (see [1]). As the solution matrix Y is positive semidefinite, a Cholesky decomposition of the Gram matrix $Y = (v_0, v_1, \ldots, v_n)^T (v_0, v_1, \ldots, v_n)$ with vectors v_i may be computed in time $\mathcal{O}(n^3)$. This is described in [11]. Observe that the diagonal elements $y_{ii} = 1$ ensure that $\| v_i \|_2 = 1$.

We separate the vectors - belonging to the subgraph or not - according to their position relative to a random hyperplane through the origin. Unfortunately, the resulting subgraph can have too many vertices, leading to an infeasible solution. But this defect is repaired by repeatedly removing a vertex with the least weighted degree until we end up with exactly k vertices. If the size of the subgraph obtained after the rounding is less than k, we include arbitrary vertices. The random experiment and the repairing step are repeated several times and finally the best output is chosen.

Algorithm SUBGRAPH

1. *Relaxation:*
   Solve the semidefinite program and compute a Cholesky decomposition of Y in order to construct solution vectors $v_0, \ldots, v_n \in B_1$.

2. *Randomized Rounding:*
   Randomly choose a unit length vector $r_t \in \mathbb{R}^{n+1}$ (to be considered as the normal of a hyperplane through the origin) and set

   $$S_t = \{1 \leq i \leq n \mid \operatorname{sgn}(v_i \cdot r_t) = \operatorname{sgn}(v_0 \cdot r_t)\}$$

   (Here sgn( ) denotes the signum function.)
3. *Repairing:*
   - If $|S_t| < k$, arbitrarily add $k - |S_t|$ vertices to the graph.
   - If $|S_t| > k$, determine a vertex $i \in S_t$ with minimum weighted degree $\sum_{j \in S_t} w_{ij}$ and remove it from S_t. Repeat this operation until S_t has k vertices. Denote the resulting vertex set by $\tilde{S}_t$.
4. *Iteration:*
   Let $T = T(\varepsilon)$ for a small $\varepsilon > 0$, repeat the steps 2 and 3 for $t = 1, \ldots, T$, and output the best solution found in one of the T runs. (T will be fixed in the analysis of the algorithm in Section 3).

We observe the following relation between the sets S_t after the rounding and $\tilde{S}_t$ after the repairing. Here we only need to consider the case $|S_t| > k$, because otherwise we may simply add arbitrary vertices increasing the weight of the subgraph.

Lemma 1. *After the removal of the vertices we have*

$$\omega(\tilde{S}_t) \geq \frac{k\,(k-1)}{|S_t|\,(|S_t| - 1)}\,\omega(S_t)\,.$$

The above inequality is tight, e.g., for a complete graph whose edges have equal weight.

Proof. If we sum up the weights of the subgraphs induced by $S_t - \{i\}$ for all $i \in S_t$, each edge is counted $|S_t| - 2$ times because it disappears when one of its endpoints is removed. So

$$\sum_{i \in S_t} \omega(S_t - \{i\}) = (|S_t| - 2)\,\omega(S_t)\,.$$

Thus, an average argument implies that after the first deletion of a vertex v with minimum weighted degree the weight of the remaining subgraph is

$$\omega(S_t - \{v\}) \geq \frac{\sum_{i \in S_t} \omega(S_t - \{i\})}{|S_t|} \geq \frac{(|S_t| - 2)}{|S_t|}\,\omega(S_t)\,.$$

The claim is then obtained by induction . $\qquad\square$

3 Analysis of the Algorithm

The analysis of the performance of our approximation algorithm is split into two parts. We first estimate the expected weight and number of vertices of the subgraph after the rounding phase. The reasoning is similiar to the MAXCUT and MAXDICUT approximation introduced by Goemans and Williamson in [10]. We refer to this article for the details of the method. In the second part, we consider the expected approximation factor after the repairing step by relating the weight of the induced subgraph and the number of vertices after the rounding in an appropriate way. Frieze and Jerrum [8] used analogous techniques for the MAXBISECTION problem. We restrict ourselves to the case $k = n/c, c > 1$. The approximation factors for $k \in \{n/2, n/3, n/4\}$ are given in the table in Section 1.

Lemma 2. *Let HS_{opt} denote the optimum of the program (HS) and SDP_{opt} the optimum of the semidefinite relaxation (SDP). For $t = 1, \ldots, T$ the subgraph induced by S_t after the rounding satisfies*

(i) $E[\omega(S_t)] \geq \beta \, SDP_{opt} \geq \beta \, HS_{opt}$

(ii) $\alpha k \leq E[|S_t|] \leq (1 - \alpha) n + \alpha k$.

Here $\alpha > 0.87856$ and $\beta > 0.79607$ are the constants Goemans and Williamson proved in the approximation algorithms for MAXCUT and MAXDICUT [10]. The derivation of the bounds closely follows the methods they applied in their analysis. For completeness we will repeat the key ideas here.

Proof. The probability that two vectors v_i and v_j are on opposite sides of the random hyperplane is proportional to the angle between those two vectors and is

$$\Pr[\operatorname{sgn}(v_i \cdot r_t) \neq \operatorname{sgn}(v_j \cdot r_t)] = \frac{\arccos(v_i \cdot v_j)}{\pi}.$$

Due to the linearity of expectation we have

$$E[\omega(S_t)] = \sum_{(i,j) \in E} w_{ij} \Pr[\operatorname{sgn}(v_i \cdot r_t) = \operatorname{sgn}(v_j \cdot r_t) = \operatorname{sgn}(v_0 \cdot r_t)].$$

In order to determine the above probability we define the following events

$$\begin{aligned}
A: &\quad \operatorname{sgn}(v_i \cdot r_t) = \operatorname{sgn}(v_j \cdot r_t) = \operatorname{sgn}(v_0 \cdot r_t) \\
B_i: &\quad \operatorname{sgn}(v_i \cdot r_t) \neq \operatorname{sgn}(v_j \cdot r_t) = \operatorname{sgn}(v_0 \cdot r_t) \\
B_j: &\quad \operatorname{sgn}(v_j \cdot r_t) \neq \operatorname{sgn}(v_i \cdot r_t) = \operatorname{sgn}(v_0 \cdot r_t) \\
B_0: &\quad \operatorname{sgn}(v_0 \cdot r_t) \neq \operatorname{sgn}(v_i \cdot r_t) = \operatorname{sgn}(v_j \cdot r_t)
\end{aligned}$$

and observe that

$$\Pr[A] + \Pr[B_i] + \Pr[B_j] + \Pr[B_0] = 1$$

and that, for instance, $\Pr[B_i] = \Pr[\operatorname{sgn}(v_j \cdot r_t) = \operatorname{sgn}(v_0 \cdot r_t)] - \Pr[A]$. Similar equations hold for $\Pr[B_j]$ and $\Pr[B_0]$. Combining these equations leads to

$$\begin{aligned}
\Pr[A] &= 1 - \frac{1}{2\pi} \Big(\arccos(v_0 \cdot v_i) + \arccos(v_0 \cdot v_j) + \arccos(v_i \cdot v_j) \Big) \\
&\geq \frac{\beta}{4} \Big(1 + v_0 \cdot v_i + v_0 \cdot v_j + v_i \cdot v_j \Big).
\end{aligned}$$

The last inequality can be verified using calculus.

Hence, we obtain $E[\omega(S_t)] \geq \beta SDP_{opt} \geq \beta H S_{opt}$ because of the relaxation. Computing the expected number of vertices in S_t we get

$$E[|S_t|] = \sum_{i=1}^{n} \Pr[\text{sgn}(v_i \cdot r_t) = \text{sgn}(v_0 \cdot r_t)]$$

$$= \sum_{i=1}^{n} \left(1 - \frac{1}{\pi} \arccos(v_0 \cdot v_i)\right)$$

$$= n - \sum_{i=1}^{n} \frac{1}{\pi} \arccos(v_0 \cdot v_i)$$

$$\leq n - \alpha \sum_{i=1}^{n} \frac{1 - v_0 \cdot v_i}{2}$$

$$= (1 - \alpha)\, n + \alpha\, k\,.$$

In the last equation we used the cardinality constraint of the relaxation (SDP). Note that

$$\frac{\pi - \arccos(v_0 \cdot v_i)}{\pi} = \frac{\arccos(-v_0 \cdot v_i)}{\pi} \geq \frac{\alpha}{2}(1 + v_0 \cdot v_i)$$

leads to the lower bound for the expected number of vertices of the subgraph. □

We continue the analysis for $k = n/c$ for some constant $c > 1$.

The main difficulty stems from the fact that so far we have only computed the expected size and weight of the subgraph, but we need to relate the size and weight to the expectations, for example by a large-deviation argument. Unfortunately, our random variables are not independent, so Chernoff-type bounds cannot be used. Fortunately, the Markov inequality already helps.

Lemma 3. *Let $\varepsilon > 0$ be some small constant.*

$$\Pr\left[|S_t| \notin [(\alpha - \varepsilon)n/c,\, n]\right] \leq p'(\varepsilon) < 1\,.$$

Proof. We apply Markov's inequality and the lower bound for the expected number of vertices

$$\Pr\left[|S_t| < (\alpha - \varepsilon)\, n/c\right] = \Pr\left[n - |S_t| > (1 - \alpha/c + \varepsilon/c)\, n\right]$$

$$\leq \frac{n - E[|S_t|]}{(1 - \alpha/c + \varepsilon/c)\, n}$$

$$\leq \frac{n - \alpha\, n/c}{(1 - \alpha/c + \varepsilon/c)\, n} =: p'(\varepsilon) < 1\,.$$

□

By repeating the rounding experiment $T' = T'(\varepsilon)$ times, we can make sure that in some run, say τ,

$$|S_\tau| \in [(\alpha - \varepsilon)\, n/c,\, n] \qquad (*)$$

with probability at least $(1 - \varepsilon)$.

Theorem 1. *For $k = n/c, c > 1$ the algorithm* SUBGRAPH *computes a subgraph S with expected weight*

$$\mathrm{E}[\omega(S)] \ge (1 - c/n)(1 - \varepsilon)^2\, \hat{r}(c)\, H S_{\mathrm{opt}}.$$

$\hat{r}(c)$ *is given by*

$$\hat{r}(c) = \beta/c^2 + \frac{(c^2 - 1)(c - 1)(1 - \varepsilon)\alpha\beta}{c^4(c - \alpha + \varepsilon) - c^2(c^2 - 1)(c - 1)(1 - \varepsilon)\alpha}$$

Proof. Remember that in the algorithm we finally choose the best subgraph of the T iterations. We define three random variables for each rounding experiment $t = 1, \ldots, T$

$$X_t = \omega(S_t), \quad Y_t = n - |S_t|, \quad Z_t = \frac{X_t}{f\, SDP_{\mathrm{opt}}} + \frac{Y_t}{n - n/c}.$$

$f > 0$ is a constant (depending on c) which we shall later fix in a suitable way. The intuition behind our definition of Z_t is judging a set S_t by its weight and its violation of the cardinality constraint.
Lemma 2 ensures

$$\mathrm{E}[Z_t] \ge \beta/f + \alpha. \qquad (1)$$

A random subgraph R with n/c vertices has expected weight

$$\mathrm{E}[\omega(R)] = \sum_{(i,j) \in E} w_{ij}\, \Pr[i \text{ and } j \in R] = \frac{1}{c^2}\omega(V).$$

Hence, $SDP_{\mathrm{opt}} \ge H S_{\mathrm{opt}} \ge \frac{1}{c^2}\omega(V)$, and

$$Z_t \le \frac{\omega(V)}{f \frac{1}{c^2}\omega(V)} + \frac{n - |S_t|}{n - n/c} \le \frac{c^2}{f} + \frac{c}{c - 1}. \qquad (2)$$

(1) and (2) imply that

$$\Pr[Z_t \le (1 - \varepsilon)(\beta/f + \alpha)] \le \frac{c^2/f + c/(c - 1) - (\beta/f + \alpha)}{c^2/f + c/(c - 1) - (1 - \varepsilon)(\beta/f + \alpha)} =: p'' < 1.$$

Repeating the rounding experiment $T'' = T''(\varepsilon)$ times we can guarantee that for $Z_\tau = \max_{1 \le t \le T''} Z_t$ the "error probability" becomes very small:

$$\Pr[Z_\tau \le (1 - \varepsilon)(\beta/f + \alpha)] \le p''^T < \varepsilon.$$

Thus, with probability $1 - \varepsilon$ we have $Z_\tau \geq (1 - \varepsilon)(\beta/f + \alpha)$.
¿From now on we may assume that this inequality holds. We may even assume that the set S_τ satisfies the condition (∗)

$$|S_\tau| \in [(\alpha - \varepsilon)\,n/c,\, n].$$

We let $X_\tau = \omega(S_\tau) = \lambda\,SDP_{\text{opt}}$ for some λ, and $|S_\tau| = \mu n$ for some $\mu \in [(\alpha - \varepsilon)/c,\, 1]$. Then,

$$Z_\tau = \frac{\lambda}{f} + \frac{n - \mu n}{n - n/c} = \frac{\lambda}{f} + \frac{c}{c-1}\,(1 - \mu)$$

$$\lambda \geq (1 - \varepsilon)(\beta + \alpha f) - \frac{fc}{c-1}\,(1 - \mu) \tag{3}$$

We split the μ-interval $[(\alpha - \varepsilon)/c,\, 1]$ into two parts
$[(\alpha - \varepsilon)/c,\, 1/c[$ and $[1/c,\, 1]$ and consider them separately.
In case of $\mu \in [(\alpha - \varepsilon)/c,\, 1/c[$, the number of vertices after the rounding is too small and no vertices have to be removed. Then

$$\omega(\tilde{S}_\tau) \geq \omega(S_\tau) = \lambda\,SDP_{\text{opt}}$$

$$\geq \left[(1 - \varepsilon)(\beta + \alpha f) - \frac{fc}{c-1}\,(1 - \mu)\right] SDP_{\text{opt}}$$

$$\geq \left[(1 - \varepsilon)(\beta + \alpha f) - \frac{f}{c-1}\,(c - \alpha + \epsilon)\right] SDP_{\text{opt}} \tag{4}$$

For $\mu \in [1/c,\, 1]$ the number of vertices is too large and some vertices have been deleted. Here we apply Lemma 1 in order to estimate the weight of the subgraph induced by $\tilde{S}_\tau$

$$\omega(\tilde{S}_\tau) \geq \frac{n/c \cdot (n/c - 1)}{\mu n \cdot (\mu n - 1)}\,\omega(S_\tau)$$

$$\geq \left(1 - \frac{c}{n}\right) \frac{\lambda}{c^2\,\mu^2}\,SDP_{\text{opt}}.$$

With the lower bound (3) for λ we can estimate $\frac{\lambda}{c^2 \mu^2}$:

$$\min_{\mu \in [1/c,1]} \frac{\lambda}{c^2\,\mu^2} \geq \min_{\mu \in [1/c,1]} \frac{(1 - \varepsilon)(\beta + \alpha f) - \frac{fc}{c-1}\,(1 - \mu)}{c^2\,\mu^2}$$

$$= \min\left\{(1 - \varepsilon)(\beta + \alpha f) - f,\; (1 - \varepsilon)\,(\beta + \alpha f)\,c^{-2}\right\}, \tag{5}$$

and the last equation follows from the fact that the minimum is attained for $\mu = 1/c$ or $\mu = 1$, as the above function has no minimum in the interior of the interval.
Comparing the factor in (4) and the first expression in (5) yields that the former is smaller than the latter. We may now choose f so that the minimum of the resulting two factors

$$\min\left\{(1 - \varepsilon)(\beta + \alpha f) - \frac{f}{c-1}\,(c - \alpha + \epsilon),\; (1 - \varepsilon)\,(\beta + \alpha f)\,c^{-2}\right\}$$

is maximized. Since the first term is a decreasing linear function of f and the second term is an increasing function, the minimum is maximized when both expressions are equal, that is for

$$f = \frac{(c^2 - 1)(c - 1)(1 - \varepsilon)\beta}{c^2(c - \alpha + \varepsilon) - (c^2 - 1)(c - 1)(1 - \varepsilon)\alpha}$$

So the following inequality holds

$$\omega(\tilde{S}_\tau) \geq (1 - \frac{c}{n})(1 - \varepsilon)\left[\frac{\beta}{c^2} + \frac{(c^2 - 1)(c - 1)(1 - \varepsilon)\alpha\beta}{c^4(c - \alpha + \varepsilon) - c^2(c^2 - 1)(c - 1)(1 - \varepsilon)\alpha}\right]SDP_{opt}$$

Observe that the weight of the subgraph produced by the algorithm is at least $\omega(\tilde{S}_\tau)$. Hence it is sufficient to compute the expectation of $\omega(\tilde{S}_\tau)$, and we get the claim of the theorem. □

The following example due to Michael Langberg shows that for $k = n/2$ the integrality ratio between SDP_{opt} and HS_{opt} is at least 2. Consider the complete graph on n vertices. A feasible vector configuration to the semidefinite program of weight approximately $n^2/4$ can be achieved by setting all vectors v_i equal to a single vector perpendicular to v_0. In that configuration each edge contributes 0.5 to the objective of the semidefinite program, yielding a total weight of $|E|/2$. On the other hand the optimal solution has weight $|E|/4$, thus the approximation factor of our algorithm for $k = n/2$ cannot be better than 0.5.

4 Conclusion

The approximation complexity of the HEAVIEST SUBGRAPH problem is not known, while the complexity of the related MAXIMUM CLIQUE problem is well-studied [13]. Any result in this direction would be of great interest.

On the positive side, better approximation algorithms relying on stronger relaxations might be devised. Feige and Goemans [6] gain a better approximation ratio for MAXDICUT by adding valid inequalities and using different rounding schemes.

References

1. F. Alizadeh. Interior point methods in semidefinite programming with applications to combinatorial optimization. *SIAM Journal on Optimization* 5(1): 13-51, 1995.
2. Y. Asahiro, K. Iwama, H. Tamaki, and T. Tokuyama. Greedily finding a dense subgraph. In *Proceedings of the 5th Scandinavian Workshop on Algorithm Theory (SWAT)*. Lecture Notes in Computer Science, 1097, pages 136-148, Springer-Verlag, 1996.
3. S. Arora, D. Karger, and M. Karpinski. Polynomial time approximation schemes for dense instances of $\mathcal{NP}$-hard problems. In *Proceedings of the 27th Annual ACM Symposium on Theory of Computing*, pages 284-293, 1995.

4. B. Chandra and M.M. Halldórsson. Facility dispersion and remote subgraphs. In *Proceedings of the 5th Scandinavian Workshop on Algorithm Theory (SWAT)*. Lecture Notes in Computer Science, 1097, pages 53-65, Springer-Verlag, 1996.

5. P. Crescenzi and V. Kann. A compendium of $\mathcal{NP}$ optimization problems. Technical report SI/RR-95/02, Dipartimento di Scienze dell'Informazione, Università di Roma "La Sapienza". The problem list is continuously updated and available as http://www.nada.kth.se/theory/problemlist.html.

6. U. Feige and M.X. Goemans. Approximating the value of two prover proof systems, with applications to MAX 2SAT and MAX DICUT. In *Proceedings of the 3rd Israel Symposium on the Theory of Computing and Systems*, pages 182-189, 1995.

7. U. Feige and M. Seltser. On the densest k-subgraph problem. Technical report, Department of Applied Mathematics and Computer Science, The Weizmann Institute, Rehovot, September 1997.

8. A. Frieze and M. Jerrum. Improved approximation algorithms for MAX k-CUT and MAX BISECTION. *Algorithmica* 18: 67-81, 1997.

9. M.X. Goemans. Mathematical programming and approximation algorithms. Lecture given at the Summer School on Approximate Solution of Hard Combinatorial Problems, Udine, September 1996.

10. M.X. Goemans and D.P. Williamson. Improved approximation algorithms for maximum cut and satisfiability problems using semidefinite programming. In *Journal of the ACM* 42(6): 1115-1145, 1995. A preliminary version has appeared in *Proceedings of the 26th Annual ACM Symposium on Theory of Computing*, pages 422-431, 1994.

11. G.H. Golub and C.F. van Loan. *Matrix Computations*. North Oxford Academic, 1986.

12. R. Hassin, S. Rubinstein and A. Tamir. Approximation algorithms for maximum dispersion. Technical report, Department of Statistics and Operations Research, Tel Aviv University, June 1997.

13. J. Håstad. Clique is hard to approximate within $n^{1-\varepsilon}$. In *Proceedings of the 37th Annual IEEE Symposium on Foundations of Computer Science*, pages 627-636, 1996.

14. G. Kortsarz and D. Peleg. On choosing a dense subgraph. In *Proceedings of the 34th Annual IEEE Symposium on Foundations of Computer Science*, pages 692-701, 1993.

15. S.S. Ravi, D.J. Rosenkrantz and G.K. Tayi. Facility dispersion problems: Heuristics and special cases. In *Proceedings of the 2nd Workshop on Algorithms and Data Structures*. Lecture Notes in Computer Science, 519, pages 355-366, Springer-Verlag, 1991.

Best Possible Approximation Algorithm for MAX SAT with Cardinality Constraint

Maxim I. Sviridenko *

Sobolev Institute of Mathematics, Russia

Abstract. In this work we consider the MAX SAT problem with the additional constraint that at most p variables have a true value. We obtain $(1 - e^{-1})$-approximation algorithm for this problem. Feige [5] proves that for the MAX SAT with cardinality constraint with clauses without negations this is the best possible performance guarantee unless $P = NP$

1 Introduction

An instance of the Maximum Satisfiability Problem (MAX SAT) is defined by a collection C of boolean clauses, where each clause is a disjunction of literals drawn from a set of variables $\{x_1, \ldots, x_n\}$. A literal is either a variable x or its negation $\bar{x}$. In addition for each clause $C_j \in C$, there is an associated nonnegative weight w_j. An optimal solution to a MAX SAT instance is an assignment of truth values to variables $x_1, \ldots, x_n$ that maximizes the sum of the weights of the satisfied clauses (i.e. clauses with at least one true literal). In this work we consider the cardinality constrained MAX SAT (CC-MAX SAT). A feasible truth assignment of this problem contains at most P true variables.

The MAX SAT is one of central problems in theoretical computer science. The best known approximation algorithm for the MAX SAT has performance guarantee slightly better than 0.77 [2]. In [8] it is shown that the MAX E3SAT, the version of the MAX SAT problem in which each clause is of length exactly three, cannot be approximated in polynomial time to within a ratio greater than 7/8, unless $P = NP$. It seems that for the general MAX 3SAT there exists an approximation algorithm with performance guarantee 7/8 [9]. The best known positive and negative results for the MAX-2SAT are 0,931 [6] and 21/22 [8], respectively. We can see that there is a gap between positive and negative results for the MAX SAT.

A class MPSAT is defined in [10] and it is proved that for all problems belonging to the MPSAT there exists an approximation scheme. Since the planar CC-MAX SAT belongs to the MPSAT (see the definition of this class in [10]) an existence of an approximation scheme for this problem follows. It's known that an existence of an approximation algorithm with perfomance guarantee better than $1 - e^{-1}$ for the CC-MAX SAT with clauses without negations implies $P = NP$ [5].

---

* Supported by the grant 97-01-00890 of the Russian Foundation for Basic Research.

In this work we present an approximation algorithm for the CC-MAX SAT with performance guarantee $1 - e^{-1}$. We use the method of randomized rounding of linear relaxation. Notice that for satisfiability problems without cardinality constraint best known algorithms (sometimes best possible) are obtained by using semidefinite programming relaxation (compare [3] and [6, 4, 9]) but for the CC-MAX SAT problem the best possible approximation is obtained via linear programming relaxation.

2 Linear relaxation and approximation algorithm

Consider the following integer program

$$\max \sum_{C_j \in C} w_j z_j, \tag{1}$$

subject to

$$\sum_{i \in I_j^+} y_i + \sum_{i \in I_j^-} (1 - y_i) \geq z_j \quad \text{for all } C_j \in C, \tag{2}$$

$$\sum_{i=1}^{n} y_i \leq P, \tag{3}$$

$$0 \leq z_j \leq 1 \quad \text{for all } C_j \in C, \tag{4}$$

$$y_i \in \{0, 1\} \quad i = 1, \ldots, n, \tag{5}$$

where I_j^+ (respectively I_j^-) denotes the set of variables appearing unnegated (respectively negated) in C_j. By associating $y_i = 1$ with x_i set true, $y_i = 0$ with x_i false, $z_j = 1$ with clause C_j satisfied, and $z_j = 0$ with clause C_j not satisfied, the integer program (1)-(5) corresponds to the CC-MAX SAT problem. The similar integer program was first used by Goemans and Williamson [3] for designing an approximation algorithm for the MAX SAT problem.

Let $M \geq 1$ be some integer constant. We define M in the next section. Consider the problem (1)-(5) with additional constraint $\sum_{i=1}^{n} y_i \leq M$. We can find an optimal solution (y_1, z_1) of this problem in polynomial time by complete enumeration. Consider the problem (1)-(5) with another additional constraint $\sum_{i=1}^{n} y_i \geq M$ and let (y_2, z_2) be an α-approximation solution of this problem. Clearly, the best of these two solutions is an α-approximation solution of the CC-MAX SAT. Consequently, without loss of generality we may consider the problem (1)-(5) with constraint $\sum_{i=1}^{n} y_i \geq M$.

For $t = M, \ldots, P$ consider now linear programs LP_t formed by replacing $y_i \in \{0, 1\}$ constraints with the constraints $0 \leq y_i \leq 1$ and by replacing (3) with the constraint

$$\sum_{i=1}^{n} y_i = t. \tag{6}$$

Let F_t^* be a value of an optimal solution of LP_t. Let k denote an index such that $F_k^* = \max_{M \leq t \leq n} F_t^*$. Since any optimal solution of the problem (1)-(5) with constraint $\sum_{i=1}^{n} y_i \geq M$ is a feasible solution of LP_t for some t, we obtain that F_k^* is an upper bound of the optimal value of this problem. We now present a randomized approximation algorithm for the CC-MAX SAT.

Description of algorithm

1. Solve the linear programs LP_t for all $t = M, \ldots, P$. Let (y^*, z^*) be an optimal solution of LP_k.

2. The second part of the algorithm consists of k independent steps. On each step algorithm chooses an index i from the set $\{1, \ldots, n\}$ at random with probability $P_i = \frac{y_i^*}{k}$. Let S denote the set of the chosen indices. Notice that $P \geq k \geq |S|$. We set $x_i = 1$ if $i \in S$ and $x_i = 0$ otherwise.

Our final algorithm consists of two steps. The first step is to solve linear programs LP_t for all $t = M, \ldots, P$. We can do it by using any known polynomial algorithm for linear programming. The second step is a derandomization the randomized part of the algorithm. We will show in the section 4 that derandomization can be done in polynomial time. In the next section we evaluate an expectation of the value of the rounded solution.

3 Analysis of algorithm

3.1 Preliminaries

In this subsection we state some technical lemmas.

Lemma 1. *The probability of realization of at least one among the events $A_1, \ldots, A_n$ is given by*

$$Pr(A_1 \cup \ldots \cup A_n) = \sum_{1 \leq i \leq n} Pr(A_i) - \sum_{1 \leq i_1 < i_2 \leq n} Pr(A_{i_1} \cap A_{i_2}) + \ldots$$

$$+ (-1)^{t-1} \sum_{1 \leq i_1 < \ldots < i_t \leq n} Pr(A_{i_1} \cap \ldots \cap A_{i_t}) + \ldots$$

Proof. see in [7], v.1, chapter 4.

Lemma 2. *The probability of realization of at least one among the events $B, A_1, \ldots, A_n$ is given by*

$$Pr(B \cup A_1 \cup \ldots \cup A_n) = Pr(B) + \sum_{1 \leq i \leq n} Pr(\bar{B} \cap A_i) - \sum_{1 \leq i_1 < i_2 \leq n} Pr(\bar{B} \cap A_{i_1} \cap A_{i_2}) + \ldots$$

$$+ (-1)^{t-1} \sum_{1 \leq i_1 < \ldots < i_t \leq n} Pr(\bar{B} \cap A_{i_1} \cap \ldots \cap A_{i_t}) + \ldots$$

Proof. The claim follows from lemma 1 and the facts

$$Pr(B \cup A_1 \cup \ldots \cup A_n) = Pr(B) + Pr(\bar{B} \cap (A_1 \cup \ldots \cup A_n)) =$$

$$= Pr(B) + Pr((\bar{B} \cap A_1) \cup \ldots \cup (\bar{B} \cap A_n)).$$

Lemma 3. *The inequalities*

$$1 - e^{-y} \le e^{-1+y}, \quad 1 - e^{-4/k}e^{-y} \le e^{-1+y} - f(k),$$

$$1 - e^{-4/k}e^{-y} - e^{-4/k}e^{-x} + e^{-x-y} \le e^{-2+x+y}$$

hold for all $y, x \in [0,1]$, $k \ge M$ where M is a sufficiently large constant independent of x, y and $\lim_{k \to +\infty} f(k) = 0$.

Proof. Let $x = e^y$, then the first inequality is equivalent to $x^2 - ex + e \ge 0$. Since $e^2 - 4e < 0$ we obtain the desired statement. Using the same argument we can prove the second inequality for sufficiently large k. We now prove the third inequality

$$1 - e^{-4/k}e^{-y} - e^{-4/k}e^{-x} + e^{-x-y} = (1 - e^{-4/k}e^{-x})(1 - e^{-4/k}e^{-y}) + f_1(k) \le$$

where $f_1(k) = e^{-x}e^{-y}(1 - e^{-8/k})$. Let $f(k) = e \cdot f_1(k)$ then we continue using the second inequality

$$\le (e^{-1+x} - f(k))e^{-1+y} + f_1(k) \le e^{-2+x+y}.$$

In the following lemma we will use the well-known inequalities $e^{-1} \ge (1 - 1/k)^k \ge e^{-1-1/k}$ for $k \ge 2$ and $e^{-a} \ge (1 - a/k)^k \ge e^{-a-a^2/k}$ for $k \ge 2a \ge 0$.

Lemma 4. *The inequality*

$$f(x, y, z) = \left(1 - \frac{x}{k}\right)^k + \left(1 - \frac{y}{k}\right)^k + \left(1 - \frac{z}{k}\right)^k - \left(1 - \frac{x+y}{k}\right)^k -$$

$$- \left(1 - \frac{y+z}{k}\right)^k - \left(1 - \frac{z+x}{k}\right)^k + \left(1 - \frac{x+y+z}{k}\right)^k > 1 - e^{-1}.$$

holds for all $x, y, z \in [0,1]$ and $k \ge M$ where M is a sufficiently large constant independent of x, y, z.

Proof. Notice that the following inequalities hold for all $x \in [0,1]$

$$e^{-x} - (1 - x/k)^k \le e^{-x}(1 - e^{-x^2/k}) \le 1 - e^{-1/k}.$$

Using the similar arguments we have

$$\lim_{k \to +\infty} \left\{ e^{-x} + e^{-y} + e^{-z} + e^{-x-y-z} - \left(1 - \frac{x}{k}\right)^k - \right.$$

$$\left. - \left(1 - \frac{y}{k}\right)^k - \left(1 - \frac{z}{k}\right)^k - \left(1 - \frac{x+y+z}{k}\right)^k \right\} = 0$$

and therefore for large k we obtain

$$f(x, y, z) \ge e^{-x} + e^{-y} + e^{-z} - e^{-x-y} - e^{-x-z} - e^{-y-z} + e^{-x-y-z} - o(1) =$$

$$= 1 - (1 - e^{-x})(1 - e^{-y})(1 - e^{-z}) - o(1) \ge 1 - (1 - e^{-1})^3 - o(1) > 0.74 > 1 - e^{-1}.$$

3.2 Evaluating of expectation

Let S denote the set of indices produced by randomized algorithm, let $f(S)$ be the value of the solution defined by the set S and let $E(f(S))$ be the expectation of $f(S)$. Now we prove our main statement.

Theorem 5.
$$F_k^* \geq E(f(S)) \geq (1 - e^{-1})F_k^*$$

Proof. Using the linearity of expectation we obtain

$$E(f(S)) = \sum_{C_j \in C} w_j Pr(z_j = 1).$$

Let $X^+ = \sum_{i \in I_j^+} y_i^*$. We consider four cases now.

Case 1 Assume that $|I_j^-| = \emptyset$. Since the steps of algorithm are independent and $X^+ \geq z_j^*$ we have

$$Pr(z_j = 1) = Pr(S \cap I_j^+ \neq \emptyset) = 1 - \left(1 - \frac{X^+}{k}\right)^k \geq$$

$$\geq 1 - \left(1 - \frac{z_j^*}{k}\right)^k \geq \left(1 - \left(1 - \frac{1}{k}\right)^k\right) z_j^*.$$

The last inequality follows from concavity of the function $g(z) = 1 - (1 - z/k)^k$ and the facts $g(0) = 0, g(1) = 1 - (1 - 1/k)^k$.

Case 2 Assume that $|I_j^-| = 1$. Let $I_j^- = \{t\}$ and $a = y_t^*$. If $X^+ > 1$, then using the argument of the previous case we obtain $Pr(z_j = 1) \geq Pr(S \cap I_j^+ \neq \emptyset) \geq 1 - e^{-1}$. Assume that $X^+ \leq 1$, then applying the lemmas 2,3 and inequality $X^+ + (1 - a) \geq z_j^*$ we have

$$Pr(z_j = 1) = Pr(S \cap I_j^+ \neq \emptyset \text{ or } t \notin S) =$$

$$= Pr(S \cap I_j^+ \neq \emptyset) + Pr(S \cap I_j^+ = \emptyset \text{ and } t \notin S) =$$

$$= 1 - \left(1 - \frac{X^+}{k}\right)^k + \left(1 - \frac{X^+ + a}{k}\right)^k \geq 1 - e^{-X^+} + e^{-X^+ - a - (X^+ + a)^2/k} \geq$$

$$\geq 1 - e^{-X^+} + e^{-4/k} e^{-X^+ - a} \geq 1 - e^{-X^+} e^{-1 + a} \geq 1 - e^{-z_j^*} \geq (1 - e^{-1}) z_j^*$$

Case 3 Assume that $|I_j^-| = 2$. Let $I_j^- = \{i_1, i_2\}, a = y_{i_1}^*$ and $b = y_{i_2}^*$. Without loss of generality assume that $X^+ \leq 1$, then using the lemmas 2, 3 and inequality $X^+ + (1 - a) + (1 - b) \geq z_j^*$ we obtain

$$Pr(z_j = 1) = Pr(S \cap I_j^+ \neq \emptyset \text{ or } i_1 \notin S \text{ or } i_2 \notin S) =$$

$$= Pr(S \cap I_j^+ \neq \emptyset) + Pr(S \cap I_j^+ = \emptyset \text{ and } i_1 \notin S) +$$

$$+ Pr(S \cap I_j^+ = \emptyset \text{ and } i_1 \notin S) - Pr(S \cap I_j^+ = \emptyset \text{ and } i_1 \notin S \text{ and } i_2 \notin S) =$$

$$= 1 - \left(1 - \frac{X^+}{k}\right)^k + \left(1 - \frac{X^+ + a}{k}\right)^k + \left(1 - \frac{X^+ + b}{k}\right)^k - \left(1 - \frac{X^+ + a + b}{k}\right)^k \geq$$

$$\geq 1 - e^{-X^+} + e^{-4/k} e^{-X^+ - a} + e^{-4/k} e^{-X^+ - b} - e^{-X^+ - a - b} \geq$$

$$\geq 1 - e^{-X^+} e^{-2 + a + b} \geq 1 - e^{-z_j^*} \geq (1 - e^{-1}) z_j^*$$

Case 4 Assume that $|I_j^-| \geq 3$. Let i_1, i_2, i_3 be arbitrary indices from the set I_j^- then applying lemmas 1,4 we have

$$Pr(z_j = 1) \geq Pr(i_1 \notin S \text{ or } i_2 \notin S \text{ or } i_3 \notin S) = Pr(i_1 \notin S) + Pr(i_2 \notin S) +$$

$$+ Pr(i_3 \notin S) - Pr(i_1 \notin S \text{ and } i_2 \notin S) - Pr(i_2 \notin S \text{ and } i_3 \notin S) -$$

$$- Pr(i_1 \notin S \text{ and } i_3 \notin S) + Pr(i_1 \notin S \text{ and } i_2 \notin S \text{ and } i_3 \notin S) =$$

$$= \left(1 - \frac{y_{i_1}^*}{k}\right)^k + \left(1 - \frac{y_{i_2}^*}{k}\right)^k + \left(1 - \frac{y_{i_3}^*}{k}\right)^k - \left(1 - \frac{y_{i_1}^* + y_{i_2}^*}{k}\right)^k -$$

$$- \left(1 - \frac{y_{i_2}^* + y_{i_3}^*}{k}\right)^k - \left(1 - \frac{y_{i_3}^* + y_{i_1}^*}{k}\right)^k + \left(1 - \frac{y_{i_1}^* + y_{i_2}^* + y_{i_3}^*}{k}\right)^k > 1 - e^{-1}.$$

4 Derandomization

In this section we apply the method of conditional expectations [1] to find an approximation truth assignment in polynomial time. The straightforward using of this method doesn't give a polynomial-time algorithm since if an instance of the CC-MAX SAT contains a clause with not constant number of negations we cannot calculate (by using lemma 2) the conditional expectations in polynomial time.

Let IP_1 be an instance of the CC-MAX SAT given by a set of clauses $C = \{C_j : j = 1, \ldots, m\}$ and a set of variables $\{x_1, \ldots, x_n\}$. Let F_k^* be the value of optimal solution of relaxation LP_k for IP_1. We define an instance IP_2 of the CC-MAX SAT by replacing each clause such that $|I_j^-| \geq 3$ by a clause $C_j' = \bar{x}_{i_1} \vee \bar{x}_{i_2} \vee \bar{x}_{i_3}$ where $i_1, i_2, i_3 \in I_j^-$.

We apply the randomized approximation algorithm with probabilities defined by optimal solution of LP_k. Let S be a solution obtained by randomized rounding, $f_1(S)$ - value of S for the problem IP_1 and $f_2(S)$ - value of S for the problem IP_2, then using the fact that $Pr(z_j = 1) > 1 - e^{-1}$ for all clauses C_j with $|I_j^-| \geq 3$, we have $E(f_2(S)) \geq (1 - e^{-1}) F_k^*$. We can derandomize this algorithm using the following procedure

Description of derandomization

Derandomized algorithm consists of k steps. On s-th step we choose an index i_s which maximizes a conditional expectation, i.e.

$$E(f_2(S)|i_1 \in S, \ldots, i_{s-1} \in S, i_s \in S) = \max_{j \in \{1, \ldots, n\}} E(f_2(S)|i_1 \in S, \ldots, i_{s-1} \in S, j \in S).$$

Since

$$\max_{j \in \{1, \ldots, n\}} E(f_2(S)|i_1 \in S, \ldots, i_{s-1} \in S, j \in S) \geq E(f_2(S)|i_1 \in S, \ldots, i_{s-1} \in S)$$

at the end of derandomization we obtain a solution $\tilde{S}$ such that $f_2(\tilde{S}) \geq E(f_2(S)) \geq (1-e^{-1})F_k^*$. Since $F_k^* \geq f_1(\tilde{S}) \geq f_2(\tilde{S})$ this solution is an $(1-e^{-1})$-approximation solution for IP_1. We can calculate a conditional expectations in polynomial time using their linearity, lemma 2 and the fact that $|I_j^-| \leq 3$ in IP_2.

References

1. Alon, N., Spencer, J.H.: Probabilistic Method. Wiley, 1992.
2. Asano, T.: Approximation algorithms for MAX SAT: Yannakakis vs. Goemans-Williamson. In Proceedings of the 5nd Israel Symposium on Theory and Computing Systems (1997) 182–189
3. Goemans, M., Williamson, D.: New 3/4-approximation algorithms for the maximum satisfiability problem. SIAM Journal on Discrete Mathematics **7** (1994) 656–666
4. Goemans M., Williamson, D.: Imroved approximation algorithms for maximum cut and satisfiability problems using semidefinite programming Journal of ACM **42** (1995) 1115–1145
5. Feige, U.: A threshold of $\ln n$ for approximating set cover. Journal of ACM (to appear)
6. Feige, U., Goemans, M.: Approximating the value of two-prover proof systems, with applications to MAX 2-SAT and MAX-DICUT. In Proceedings of the 3nd Israel Symposium on Theory and Computing Systems (1995) 182–189
7. Feller, W.: An Introduction to Probability Theory and Its Applications. John Wiley & Sons New York (1968)
8. Hastad, J.: Some optimal inapproximability results. In Proceedings of the 28 Annual ACM Symp. on Theory of Computing (1996) 1–10
9. Karloff, H., Zwick, U.: A 7/8-approximation algorithm for MAX 3SAT? In Proceedings of the 38th FOCS (1997) 406–415
10. Khanna, S., Motwani, R.: Towards syntactic characterization of PTAS. In Proceedings of the 28 Annual ACM Symp. on Theory of Computing (1996) 329–337

Author Index

Lecture Notes in Computer Science

For information about Vols. 1–1359

please contact your bookseller or Springer-Verlag

Vol. 1396: E. Okamoto, G. Davida, M. Mambo (Eds.), Information Security. Proceedings, 1997. XII, 357 pages. 1998.

Vol. 1397: H. de Swart (Ed.), Automated Reasoning with Analytic Tableaux and Related Methods. Proceedings, 1998. X, 325 pages. 1998. (Subseries LNAI).

Vol. 1398: C. Nédellec, C. Rouveirol (Eds.), Machine Learning: ECML-98. Proceedings, 1998. XII, 420 pages. 1998. (Subseries LNAI).

Vol. 1399: O. Etzion, S. Jajodia, S. Sripada (Eds.), Temporal Databases: Research and Practice. X, 429 pages. 1998.

Vol. 1400: M. Lenz, B. Bartsch-Spörl, H.-D. Burkhard, S. Wess (Eds.), Case-Based Reasoning Technology. XVIII, 405 pages. 1998. (Subseries LNAI).

Vol. 1401: P. Sloot, M. Bubak, B. Hertzberger (Eds.), High-Performance Computing and Networking. Proceedings, 1998. XX, 1309 pages. 1998.

Vol. 1402: W. Lamersdorf, M. Merz (Eds.), Trends in Distributed Systems for Electronic Commerce. Proceedings, 1998. XII, 255 pages. 1998.

Vol. 1403: K. Nyberg (Ed.), Advances in Cryptology – EUROCRYPT '98. Proceedings, 1998. X, 607 pages. 1998.

Vol. 1404: C. Freksa, C. Habel. K.F. Wender (Eds.), Spatial Cognition. VIII, 491 pages. 1998. (Subseries LNAI).

Vol. 1405: S.M. Embury, N.J. Fiddian, W.A. Gray, A.C. Jones (Eds.), Advances in Databases. Proceedings, 1998. XII, 183 pages. 1998.

Vol. 1406: H. Burkhardt, B. Neumann (Eds.), Computer Vision – ECCV'98. Vol. I. Proceedings, 1998. XVI, 927 pages. 1998.

Vol. 1407: H. Burkhardt, B. Neumann (Eds.), Computer Vision – ECCV'98. Vol. II. Proceedings, 1998. XVI, 881 pages. 1998.

Vol. 1409: T. Schaub, The Automation of Reasoning with Incomplete Information. XI, 159 pages. 1998. (Subseries LNAI).

Vol. 1411: L. Asplund (Ed.), Reliable Software Technologies – Ada-Europe. Proceedings, 1998. XI, 297 pages. 1998.

Vol. 1412: R.E. Bixby, E.A. Boyd, R.Z. Ríos-Mercado (Eds.), Integer Programming and Combinatorial Optimization. Proceedings, 1998. IX, 437 pages. 1998.

Vol. 1413: B. Pernici, C. Thanos (Eds.), Advanced Information Systems Engineering. Proceedings, 1998. X, 423 pages. 1998.

Vol. 1414: M. Nielsen, W. Thomas (Eds.), Computer Science Logic. Selected Papers, 1997. VIII, 511 pages. 1998.

Vol. 1415: J. Mira, A.P. del Pobil, M.Ali (Eds.), Methodology and Tools in Knowledge-Based Systems. Vol. I. Proceedings, 1998. XXIV, 887 pages. 1998. (Subseries LNAI).

Vol. 1416: A.P. del Pobil, J. Mira, M.Ali (Eds.), Tasks and Methods in Applied Artificial Intelligence. Vol.II. Proceedings, 1998. XXIII, 943 pages. 1998. (Subseries LNAI).

Vol. 1417: S. Yalamanchili, J. Duato (Eds.), Parallel Computer Routing and Communication. Proceedings, 1997. XII, 309 pages. 1998.

Vol. 1418: R. Mercer, E. Neufeld (Eds.), Advances in Artificial Intelligence. Proceedings, 1998. XII, 467 pages. 1998. (Subseries LNAI).

Vol. 1420: J. Desel, M. Silva (Eds.), Application and Theory of Petri Nets 1998. Proceedings, 1998. VIII, 385 pages. 1998.

Vol. 1421: C. Kirchner, H. Kirchner (Eds.), Automated Deduction – CADE-15. Proceedings, 1998. XIV, 443 pages. 1998. (Subseries LNAI).

Vol. 1422: J. Jeuring (Ed.), Mathematics of Program Construction. Proceedings, 1998. X, 383 pages. 1998.

Vol. 1423: J.P. Buhler (Ed.), Algorithmic Number Theory. Proceedings, 1998. X, 640 pages. 1998.

Vol. 1424: L. Polkowski, A. Skowron (Eds.), Rough Sets and Current Trends in Computing. Proceedings, 1998. XIII, 626 pages. 1998. (Subseries LNAI).

Vol. 1425: D. Hutchison, R. Schäfer (Eds.), Multimedia Applications, Services and Techniques – ECMAST'98. Proceedings, 1998. XVI, 532 pages. 1998.

Vol. 1427: A.J. Hu, M.Y. Vardi (Eds.), Computer Aided Verification. Proceedings, 1998. IX, 552 pages. 1998.

Vol. 1430: S. Trigila, A. Mullery, M. Campolargo, H. Vanderstraeten, M. Mampaey (Eds.), Intelligence in Services and Networks: Technology for Ubiquitous Telecom Services. Proceedings, 1998. XII, 550 pages. 1998.

Vol. 1431: H. Imai, Y. Zheng (Eds.), Public Key Cryptography. Proceedings, 1998. XI, 263 pages. 1998.

Vol. 1432: S. Arnborg, L. Ivansson (Eds.), Algorithm Theory – SWAT '98. Proceedings, 1998. IX, 347 pages. 1998.

Vol. 1433: V. Honovar, G. Slutzki (Eds.), Grammatical Inference. Proceedings, 1998. X, 271 pages. 1998. (Subseries LNAI).

Vol. 1435: M. Klusch, G. Weiß (Eds.), Cooperative Information Agents II. Proceedings, 1998. IX, 307 pages. 1998. (Subseries LNAI).

Vol. 1436: D. Wood, S. Yu (Eds.), Automata Implementation. Proceedings, 1997. VIII, 253 pages. 1998.

Vol. 1437: S. Albayrak, F.J. Garijo (Eds.), Intelligent Agents for Telecommunication Applications. Proceedings, 1998. XII, 251 pages. 1998. (Subseries LNAI).

Vol. 1438: C. Boyd, E. Dawson (Eds.), Information Security and Privacy. Proceedings, 1998. XI, 423 pages. 1998.

Vol. 1439: B. Magnusson (Ed.), Software Configuration Management. Proceedings, 1998. X, 207 pages. 1998.

Vol. 1441: M. Pagnucco, W. Wobcke, C. Zhang (Eds.), Agents and Multi-Agent Systems. Proceedings, 1997. XII, 241 pages. 1998. (Subseries LNAI).

Vol. 1444: K. Jansen, J. Rolim (Eds.), Approximation Algorithms for Combinatorial Optimization. Proceedings, 1998. VIII, 201 pages. 1998.

Vol. 1445: E. Jul (Ed.), ECOOP'98 – Object-Oriented Programming. Proceedings, 1998. XII, 635 pages. 1998.

Vol. 1446: D. Page (Ed.), Inductive Logic Programming. Proceedings, 1998. VIII, 301 pages. 1998. (Subseries LNAI).

Vol. 1448: M. Farach-Colton (Ed.), Combinatorial Pattern Matching. Proceedings, 1998. VIII, 251 pages. 1998.